# The Poetry of Physics
### and
# The Physics of Poetry

# The Poetry of Physics
and
# The Physics of Poetry

## Robert K Logan

Department of Physics
University of Toronto, Canada
and
Strategic Innovation Lab (sLab)
Ontario College of Art and Design

 **World Scientific**

NEW JERSEY · LONDON · SINGAPORE · BEIJING · SHANGHAI · HONG KONG · TAIPEI · CHENNAI

*Published by*

World Scientific Publishing Co. Pte. Ltd.

5 Toh Tuck Link, Singapore 596224

*USA office:* 27 Warren Street, Suite 401-402, Hackensack, NJ 07601

*UK office:* 57 Shelton Street, Covent Garden, London WC2H 9HE

**British Library Cataloguing-in-Publication Data**
A catalogue record for this book is available from the British Library.

ISBN-13 978-981-4295-92-5
ISBN-10 981-4295-92-2
ISBN-13 978-981-4295-93-2 (pbk)
ISBN-10 981-4295-93-0 (pbk)

Printed in Singapore.

*This book is dedicated to my students, past and present, of the Poetry of Physics as well as Rebecca F., Haviva, Jacob, Joshua, Rebecca L., Natalie, David, Renee, John, Paul and Mana.*

# Contents

# Chapter 1

# To My Readers

Poets say science takes away from the beauty of the stars — mere globs of gas atoms. I, too, can see the stars on a desert night, and feel them. But do I see less or more? – Richard Feynman

There is poetry in physics and physics in poetry. This book is the product of a course I taught at the University of Toronto starting in 1971 and which I am still teaching at the date of this publication. The course was entitled the Poetry of Physics and the Physics of Poetry. The course was first taught at University College of the University of Toronto and then switched to New College where I also organized a series of seminars on future studies known as the Club of Gnu. After a short recess the course then became a Department of Physics course and was offered as a seminar course for first year students. The purpose of the course that I have now taught for the past 38 years was to introduce the ideas of physics to humanities and arts student who would not otherwise be exposed to these ideas and to try to address the alienation to science that so many of the lay public feel, which is a characteristic of our times. By studying physics without math you, the reader, will encounter the poetry of physics. We will also examine some of the impacts of physics on the humanities and the arts. This is the physics of poetry.

The alienation represented by the gap between the sciences and the humanities is frequently referred to as the two cultures. There are two factors contributing to this alienation; one is the basic lack of understanding of the actual subject matter of science and the other a misunderstanding of the role science plays in our society. Although the fear of science is quite pervasive I believe there are many people interested in leaning about physics. The word "physics" is derived from the Greek word phusis, which means nature. Those that are curious about

1

the "nature" of the world in which they live should, therefore, want to study physics.

This unfortunately, is not always the case, due in part to the fact that historically physics has been taught in a manner, which alienates most students. This has been accomplished by teaching physics mathematically, which has resulted in more confusion than elucidation for many. Also because the easiest way to examine students and assign grades is to ask quantitative questions, there has been a tendency to teach the formulae of physics rather than the concepts.

This book attempts to remedy this classical situation by communicating the ideas of physics to the reader without relying on mathematics. Mathematical formulae are used, but only after the concepts have been carefully explained. The math will be purely supplementary and none of the material developed later in the book will depend on these formulae. The role of a mathematical equation in physics is also described. To repeat the mathematics is purely supplementary. This book is written explicitly for the people who have difficulty with the mathematics but wish to understand their physical universe. Although all fields of physics are covered the reader will find a bit more emphasis on the modern physics that emerged in the beginning of the 20th century with quantum mechanics and Einstein's theory of relativity. The reason for this is that this physics is less intuitive than classical physics and hence requires more of an explanation.

A second aim of the book is to understand the nature of science and the role it plays in shaping both our thinking and the structure of our society. We live in times when many of the decisions in our society are made by professionals claiming scientific expertise. Science is the password today with those who study social and political problems. They label themselves social scientists and political scientists. It is, therefore, vital to the survival of our society that there exists a general understanding of what science is, what it can do and perhaps most importantly what it cannot do. I have therefore, made an attempt to shed as much light on the scientific process as possible. We will demonstrate that science unlike mathematics cannot prove the truth of its propositions but that it must constantly test its hypotheses.

To restore the perspective of what science is really about we will examine science as a language, a way of describing the world we live in. To this end we will briefly examine the origin and the evolution of language to reveal how the language of science emerged. We will show that the spirit of trying to describe the physical world we live in is universal and can be traced back to preliterate societies and their oral creation myths. It was with writing that the first signs of scientific thinking began to emerge. We will also explain how alphabetic writing influenced the development of abstract science in the West despite the fact that most of technology emerged in China. We will also document the contributions to science by other non-European cultures once again demonstrating the universality of scientific thinking. Hindu mathematicians invented zero and Arab mathematicians transmitted it to Europe providing the mathematical tools for modern science. Arab scientists and scholars contributed to the scientific revolution in Renaissance Europe through their accomplishments in algebra, chemistry and medicine.

Finally, I hope that through this book I will be able to share with the reader the mystical feelings that once characterized our response to our physical environment. Unfortunately, there has arisen in many people's minds a division between the mystical and the scientific. For those in tune with their universe there is no division. In fact quite the opposite is true as these words of Albert Einstein reveal:

> The most beautiful and most profound emotion we can experience is the sensation of the mystical. It is the power of all true science. He to whom this emotion is a stranger, who can no longer stand, rapt in awe, is as good as dead. That deeply emotional conviction of the presence of a superior reasoning power, which is revealed in the incomprehensible universe, forms my idea of God.

Hopefully, the beauty of the concepts of physics will be conveyed so that the reader will come to appreciate the poetry of physics.

In addition to the poetry of physics we will also examine in this book the physics of poetry by which we mean the ways in which physics has influenced the development of poetry and all of the humanities including painting, music, literature and all of the fine arts. Interspersed within our description of the evolution of science we will examine how the arts were influenced by science and vice-versa how the arts and humanities influenced science. There will be more of a focus on poetry because like science it is pithy and it will be easy to demonstrate how science impacted on this art form by quoting from poets ranging from the poetry of creation myths to the poetry of modern times.

This book was written for first year students at the University of Toronto and as such it can be used as a textbook for an introductory non-mathematical two-term physics or science course. It is also written in such a way as to appeal to the general reader. For instructors wishing to use this book as a textbook I have provided some suggestions in Chapter 29 on topics for essay assignments or for classroom discussions.

I would welcome comments or questions from my readers via email at logan@physics.utoronto.ca.

# Chapter 2

# The Origin of Physics

What is physics? One way to answer this question is to describe physics as the study of motion, energy, heat, waves, sound, light, electricity, magnetism, matter, atoms, molecules, and nuclei. This description, aside from sounding like the table of contents of a high school physics textbook, does not really specify the nature of physics. Physics is not just the study of the natural phenomena listed above but it is also a process; a process, which has two distinguishable aspects.

The first of these is simply the acquisition of knowledge of our physical environment. The second, and perhaps more interesting, is the creation of a worldview, which provides a framework for understanding the significance of this information. These two activities are by no means independent of each other. One requires a worldview to acquire new knowledge and vice versa one needs knowledge with which to create a worldview. But how does this process begin? Which comes first, the knowledge or the worldview?

In my opinion, these two processes arise together, each creating the conditions for the other. This is analogous to a present day theory concerning the existence of elementary particles. According to the bootstrap theory, the so-called elementary particles such as protons, neutrons, and mesons are actually not elementary at all but rather they are composites of each other and they bootstrap each other into existence. But, we are getting ahead of our story. We shall wait till later to discuss the bootstrap theory of elementary particles. For now, it is useful to recognize the two aspects of the process of physics described above. Another way to describe the relationship between "the gathering of facts" and "the building of a framework for the facts" is in term of autocatalysis. Autocatalysis occurs when a group of chemicals catalyze each other's production. Stuart Kauffman has argued that life began

5

as the autocatalysis of a large set of organic chemicals that were able
to reproduce themselves.

The study of physics is generally recognized to be quite old but there
are differences of opinion as to how old. Some would argue that physics
began in Western Europe during the Renaissance with the work of
Copernicus, Galileo, Kepler, and Newton. Others would trace the
beginnings back to the early Greeks and credit the Ionian, Thales, with
being the world's first physicist. Still others would cite the even older
cultures of Mesopotamia, Egypt and China. For me, physics or the study
of nature is much older having begun with the first humans.

Humans became scientists for the sake of their own survival. The
very first toolmakers were scientists. They discovered that certain
objects in their physical environment were useful for performing
certain tasks. Having learned this they went on to improve on these
found objects first by selecting objects more suitable for the task
involved and later actually altering the materials they found to produce
manufactured tools. This activity is usually referred to as the creation
of technology. But the type of reasoning involved in this process is
typical of the scientific method, which begins with observations of
nature and moves on to generalizations or hypotheses that are tested.
For early humans, the generalizations that were made were not in the
form of theoretical laws but rather as useful tools. This is exemplified by
the achievement of tools for hunting and gathering, pastoralism and
agriculture and the use of herbs for rudimentary medicine. All of these
activities required a sophisticated level of scientific reasoning. One might
dispute this conclusion by claiming that these achievements were
technological and not scientific. We usually refer to the acquisition of
basic information as science and its application to practical problems as
technology. While this distinction is useful when considering our highly
specialized world — its usefulness when applied to early human culture
is perhaps not as great. A technological achievement presupposes the
scientific achievement upon which it is based. The merging of the
technological and scientific achievements of early humans has obscured
our appreciation of their scientific capacity.

Primitive science, rooted totally in practical application also differs
from modern science and even ancient Greek science in that it is
less abstract. Astronomy was perhaps our first abstract scientific
accomplishment, even though it was motivated by the needs of farmers
who had to determine the best time to plant and harvest their crops.

An example of the sophistication of early astronomy is the megalithic structure of Stonehenge built in approximately 2000 B.C. in England, constructed with great effort using heavy rocks weighing up to 50 tons. G.S. Hawkins (1988) in his fascinating book Stonehenge Decoded concludes that Stonehenge was not merely a temple as originally thought but actually an astronomical observatory capable of predicting accurately lunar eclipses as well as the seasonal equinoxes. One cannot help but be impressed when one realizes that the builders of Stonehenge had determined a 56-year cycle of lunar eclipses.

In his book The Savage Mind, Levi-Strauss (1960) reveals another aspect of the scientific sophistication of so-called primitive human cultures whose knowledge of plants rivals that of modern botanists. In fact, Levi-Strauss points out that contemporary botanists discovered a number of errors in their classification scheme based on the work of Linneaus by studying the classification scheme or certain South American Indians.

The examples of early scientific activity so far discussed have centered about the fact gathering aspect of physics. Evidence of interest in the other aspect of physics, namely the creation of a worldview, is documented by the mythology of primitive people. All of the peoples of the world have a section of their mythology devoted to the creation of the universe. This is a manifestation of the universal drive of all cultures to understand the nature of the world they inhabit. A collection of creation myths assembled by Charles Long (2003) in his book Alpha illustrates the diversity of explanations provided by primitive cultures to understand the existence of the universe. Amidst this diversity a pattern emerges, however, which enables one to categorize the various creation myths into different classes of explanations. One of the interesting aspects of Long's collection is that within a single class of explanations one finds specific examples from diverse geographical locations around the globe attesting to the universality of human thought. One also finds that within a single cultural milieu more than one type of explanation is employed in their mythology.

Perhaps the oldest group of emergence myths is the one in which the Earth arises from a Mother Earth Goddess as represented by mythology of North American Indians, Islanders of the South Pacific, and the people living on the north eastern frontier of India. In another set of myths the world arises from the sexual union of a father sky god and a mother Earth goddess. Examples of this form are found in the mythology of ancient Egypt, Greece, India, Babylonia, Polynesia and North America. Other classes of myths include creation by an earth diver, creation from a cosmic egg, creation from chaos, and creation from nothing. In the earth diver myths an animal or god dives into a body of water to retrieve a tiny particle of earth, which then expands to become the world. The cosmic egg myths tell of an egg, usually golden, which appears at the first moment of the universe. The egg breaks open and the events of the universe unfold. In one version the upper part of the eggshell becomes the heavens and the lower part, the Earth. At the beginning of the creation from chaos myths there is disorder or chaos sometimes depicted as water from which a creator creates the universe. Finally, in the creation from nothing myths, which are closely related to the chaos myths, the original starting point of the universe is a void. The best-known example of this group to Western readers, of course, is Genesis, where we read, "In the beginning, God created the heavens and the Earth. The Earth was without form and void and darkness was upon the face of the deep". Other examples of the creation from nothing myth are found among the ancient Greeks, the Australian aborigines, the Zuni Indians of the southwest United States, the Maori of New Zealand, the Mayans of ancient Mexico, and the Hindu thinkers of ancient India.

Having briefly surveyed the various types of creation myths, let us turn to an example of the earliest type and retell the story of Kujum-Chantu, an emergence myth told by the people who live along the northeast frontier of India:

> At first Kujum-Chantu, the Earth, was like a human being; she had a head, and arms and legs, and an enormous fat belly. The original human beings lived on the surface of her belly. One day it occurred to Kujum-Chantu that if she ever got up and walked about, everyone would fall off and be killed, so she herself died of her own accord. Her head became the snow-covered mountains; the bones of her back turned into smaller hills. Her chest was the valley where the Apa-Tanis live. From her neck came the north country of the Tagins. Her buttocks turned into the Assam plain. For just as the buttocks are full of fat, Assam has fat rich soil. Kujum-Chantu's eyes became the Sun and Moon. From her mouth was born Kujum-Popi, who sent the Sun and Moon to shine in the sky.

The story of Kujum-Chantu attempts a coherent explanation of both the creation of the world and the nature of its physical features and as such it may legitimately be regarded as a scientific hypothesis. Let us compare it with a modern day hypothesis to explain the existence and the nature of the Earth. According to the modern theory, the Earth, the other planets and the Sun were formed together from the same cosmic dust, which explains the various physical features of the Earth such as its molten iron core, its chemical composition and the nature of its physical features. Although the story of Kujum-Chantu may be considered a hypothesis in the loosest sense of the word, it must be conceded that the modern day theory does a better job of explaining the presently known facts about the Earth and as such is considered a more satisfactory scientific theory. It should also be pointed out, however, that there does not exist a set of truly objective criterion for choosing one hypothesis over another.

From our modern scientific point of view we prefer the second theory because it explains more facts. From the point of view of the member of the culture, which worships Kujum-Chantu their story probably gives them a deeper appreciation of the world. Contrary to popular belief there is no scientific manner for arbitrating between two rival scientific theories. Believe it or not, the choice is made on the basis of which theory is most satisfying on human grounds. Copernicus' Sun centered theory of the solar system was preferred at first by its proponents on aesthetic grounds. We shall return to this question when we discuss T.S. Kuhn's (1972) excellent book, *The Structure of Scientific Revolutions* in Chapter 16.

Treating the story of Kujum-Chantu and the modern theory of the creation of the solar system as equivalent theories for the purposes of illustration is perhaps a bit of an exaggeration on my part. The two rival pictures actually differ in a very crucial manner, which actually disqualifies the story of Kujum-Chantu as a bonafide scientific hypothesis. The difference is that the Kujum-Chantu hypothesis does not make any predictions whereas the modern science hypothesis makes a number of predictions, such as the relative chemical composition of the various planets including the Earth and the Sun. A theory, which makes no prediction, is merely an ad hoc (after the facts) explanation of facts, which cannot be tested. A theory, which has the possibility of being proven wrong because of its predictions, but, nevertheless, continues to explain new facts, inspires confidence in its validity. Although there is no objective criterion for choosing theories, the predictive capabilities of a hypothesis have historically provided the mechanism of choice. The best argument that can be made to justify this criterion is that it works. Its adoption has lead to the incredible wealth of knowledge that we now possess.

Science cannot prove that a hypothesis is correct. It can only verify that the hypothesis explains all observed facts and has passed all experimental tests of its validity. Only mathematics can prove that a proposition is true but that proof has to be based on some axioms that are assumed to be obviously or self-evidently true. Karl Popper (1959 and 1979), was annoyed by those Marxists and Freudians, who always wriggled out of any contradiction between their predictions and observations with some ad hoc explanation. He proposed that for a proposition to be considered a hypothesis of science it had to be falsifiable. Using Popper's criteria as an axiom I (Logan 2003) was able to prove that science cannot prove that a proposition is true. If one proved a proposition was true then it could not be falsified and therefore according to Popper's criteria it could not be considered a scientific proposition. Therefore science cannot prove the truth of one of its propositions. This is the difference between science and mathematics. Science studies the real world and mathematics makes up its own world. Scientists, however, make use of mathematics to study and describe the real world.

The two aspects of physics involving the acquisition of information and the creation of a world picture have one feature in common — they both provide us with a degree of comfort and security. The first aspect contributes to our material security. Knowledge of the physical environment and how it responds to our actions is essential to planning one's affairs. It is from this fact acquiring aspect of physics that technology arises. It is from the second or synthesizing aspect of physics, however, that we derive the psychological comforts that accrue from the possession of a worldview. The possession of a worldview is usually associated with philosophy and religion and not physics. This, unfortunately, is our modern predicament. It should be recalled that for preliterate cultures physics, philosophy and religion were integrated. The same was true for Greek culture. Perhaps the enormous mismanagement of our material resources and our environment, which characterizes our times, could be eliminated if we could once again integrate philosophy, religion and physics.

# Chapter 3

# Ancient Science of Mesopotamia, Egypt and China

The first signs of science began to emerge in ancient Mesopotamia, Egypt and China as these agricultural political economies began to develop technologies to enhance their economies. The knowledge or science that they began to acquire was not systematized and no attempt was made to relate the different discoveries that they were making into a theoretical understanding of their universe. This development had to await the emergence of Greek philosophy, which we will examine in Chapter 4. Nevertheless the three ancient cultures that we will examine in this chapter began to acquire knowledge of mathematics, astronomy, chemistry, botany, zoology, medicine and mechanics and in the case of the Chinese all of these plus magnetism and clockworks.

## Mesopotamian Writing and Science

The culture of Mesopotamia refers to the cultures that developed in the valleys of the Tigris and Euphrates river systems. The first culture there was the Sumerian, a non-Semitic speaking people whose origin and language remains a mystery to this day. They were conquered by a Semitic-speaking people, the Akkadians, who are known more commonly as the Babylonian. The Sumerians were the first culture to have invented writing and a mathematical notation. It is believed that the idea of writing spread to China in the East and Egypt in the West and from there to all the cultures of the Old World. Writing was invented independently in the New World by the Mayans. It spread from there to a few other cultures before the arrival of the Europeans. It is possible that the Inca also had a notation systems based on knots tied in ropes, known as quipas but it never flowered into a true writing system as far as we know. The existence of a writing system it seems is essential for a culture

11

to engage in scientific activity. It was only those cultures that possessed a writing system and a system for numerical notation that ever engaged in formal scientific activity.

Not only did writing first emerge in Sumer it was also here that the first formal schools were organized to teach the 3R's, the mysterious skills of reading, writing and arithmetic. It was in these scribal schools that the first primitive forms of science appeared. The major aim of the scribal school quite naturally was professional training to satisfy the economic and administrative needs of temple and palace bureaucracies. "However, in the course of its growth and development, and particularly as a result of the ever widening curriculum, the school came to be the center of culture and learning in Sumer. Within its walls flourished the scholar-scientist, the man who studied whatever theological, botanical, zoological, mineralogical, geographical, mathematical, grammatical, and linguistic knowledge was current in his day, and who in some cases added to the knowledge (Kramer 1959, p. 2)."

Writing and mathematical notation emerged simultaneously in Sumer in 3100 B.C. as was shown by the work of Denise Schmandt-Besserat (1978, 1980, 1981 & 1992). She showed how clay accounting tokens used throughout the Middle East circa 8000 to 3000 B.C. were the forerunners of writing and mathematical notation. Manual labourers in Sumer were divided into two groups, farmers and irrigation workers. The farmers had to pay tributes to the priests in the form of agricultural commodities that were redistributed to the irrigation workers. The farmers were given clay tokens as receipts for their tributes. These tokens two to three centimeters in size and each with a unique shape to represent a different agricultural commodity were sealed inside of opaque clay envelopes. This system developed because of an information overload; it was impossible using spoken language to remember all of the tributes that the priests received. Some brilliant civil servant/priest suggested that before placing the tokens inside the clay envelopes they should impress the token on the surface of the clay envelope while it was still wet so they would not need to break open the envelope each time they wanted to know what was inside. Within fifty years of this development they did away with storing the tokens inside the envelopes and just pressed the tokens on the surface of the envelope without sealing the tokens inside. The impressed envelopes became tablets.

The next development occurred within the city-state of Sumer where they dealt with large quantities and hence a new information overload arose. They developed a system where the token for a ban, a large measure of wheat (a bushel), was used to represent the abstract number ten and a token for the bariga, a small measure of wheat (a peck), was used to represent the abstract number one. If they wanted to record a transaction involving 13 lambs what they did instead of pressing the lamb token into a tablet 13 times was to press the ban token into the wet clay once, the bariga token three times and then they etched the shape that the lamb token into the wet clay with a stylus and this was read as 13 lambs. The reason they etched the shape of lamb token into the clay rather than pressing the lamb token into the clay is that the tablet would be read as one ban of wheat, three barigas of wheat and one lamb instead of 13 lambs. These etched outlines of tokens became the first written words and the impressed ban and bariga tokens the first notated numbers.

So writing and math started out as a back of the envelope doodle. They were not the invention of writers or mathematicians but humble priests/civil servants who were record keepers. Once reading and writing emerged schools had to be organized to teach these new skills because one cannot learn how to read, write and do arithmetic by watching others do it. It is not the automatic learning that takes place when we learn to talk as young children by listening to our parents and other caregivers speak. The first schools were rectangular rooms that held 30 to 40 students sitting on benches and one teacher at the head of the class (Kramer 1956). The lessons were in reading, writing and arithmetic, a tradition that has lasted 5000 years and will probably continue as long as humans walk upon this Earth.

To prepare their lessons teachers created lists of similar objects like trees, animals, fish, kings, and rivers. These teachers subsequently became scholars. The teacher who prepared the lists of trees headed the botany department and the one who created the list of kings became the political science expert. With scholarship another information overload developed from all the scholars, which was resolved with the emergence of science, a form of organized knowledge beginning around 2000 B.C.

Science emerged as organized knowledge to deal with the information overload created by teacher/scholars. The methods and findings of science are expressed in the languages of writing and mathematics, but science may be regarded as a separate form of language because it has a unique way of systematically processing, storing, retrieving, and organizing information, which is quite different from either writing or mathematics.

The elements of universality, abstraction, and classification that became part and parcel of Babylonian thinking under the influence of phonetic writing subliminally promoted a spirit of scientific investigation, which manifested itself in the scribal schools. The major aim of the scribal school quite naturally was professional training to satisfy the economic and administrative needs of temple and palace bureaucracies.

> However, in the course of its growth and development, and particularly as a result of the ever-widening curriculum, the school came to be the center of culture and learning in Sumer. Within its walls flourished the scholar-scientist, the man who studied whatever theological, botanical, zoological, mineralogical, geographical, mathematical, grammatical, and linguistic knowledge was current in his day, and who in some cases added to the knowledge (Kramer 1959, p. 2).

During the reign of Hammurabi both the writing system and the legal system in the form of the Hammurabic code were regularized and reformed. The writing system that was phonetic and based on a syllabary was reduced to 60 symbols representing the 60 syllables in terms of which all of the words of their spoken language could be represented. Weights and measures were also standardized. These developments were not coincidental. These reforms promoted the paradigms of abstraction, classification, and universality and thus encouraged the development of scientific thinking.

The next two centuries after these reforms represent the first great scientific age of mankind. A new spirit of empiricism and scholarly interest in astronomy, magic, philology, lexicography, and mathematics arose. A primitive place number system was invented as well as algorithms for arithmetic calculations. Mathematical tables were created to simplify calculations. Achievements in algebra included solutions of quadratic equations. Lists of stars and constellations were compiled and the movements of the planets were charted. The scholars of the Hammurabic era "showed such taste and talent for collecting and systematizing all recognized knowledge that Mesopotamian learning nearly stagnated for a thousand years thereafter. ...We find a pervasive idea of order and system in the universe, resulting in large part from the tremendous effort devoted to the systematization of knowledge (Albright 1957, pp. 197–99)."

The Mesopotamians' spirit of order and system is reflected in their cosmology or concept of the universe (Kramer 1959, pp. 77–79). The Babylonian universe, an-ki, is divided into two major components: the heaven (an) and the earth (ki), which emerged from and remain fixed and immovable in a boundless sea, Nammu. Nammu acts as the "first cause" or "prime mover" of the universe. Between heaven and earth there moves Lil, a divine wind (also air, breath, or spirit) from which the luminous bodies (the sun, moon, planets, and stars) arose. The order of creation is as follows: 1) the universe, an-ki (heaven-earth), emerges from the boundless sea Nammu; 2) it separates into heaven and earth; 3) Lil then arises between heaven and earth; 4) from which the heavenly bodies emerge; 5) followed finally by the creation of plants, animals, and human beings. The order of creation found in this cosmogony closely parallels the story of creation found in the Bible in the book of Genesis.

Although Mesopotamian cosmology and cosmogony was polytheistic in nature, there nevertheless evolved some rather abstract notions of the deities that created and controlled the universe. All the elements of the cosmos were attributed to four gods who controlled the heavens, earth, sea, and air. "Each of these anthropomorphic but superhuman beings was deemed to be in charge of a particular component of the universe and to guide its activities in accordance with established rules and regulations (ibid., p. 78)." These four spheres of influence correspond to the four elements of fire, air, water and earth from which the Greeks composed their universe more than a thousand years later.

While Mesopotamian cosmology contains mythic elements, the core of its world picture is based on empirical observations of the natural environment including the heavens. Systematic astronomical observations were not part of the Sumerian tradition but were begun by the Akkadians, worshippers of the sun god Shamash. Their observations were somewhat crude (Neugebauer 1969, p. 97) and it was only with the flowering of the Assyrian empire in approximately 700 B.C. that accurate quantitative measurements were made (ibid., p. 101). Tablets recording these observations have been used to date the chronology of the Hammurabic period (ibid., p. 100). Part of the motivation for these observations was what we could term scientific and part astrological, though the Babylonians made no distinction between science and astrology. Observations made for the purpose of divination served science as well, and paradoxically, vice versa.

Sumerian and Babylonian mathematical tables provide further evidence for the development of scientific thinking in Mesopotamia. These tables were combined with tables of weights and measures indicating that they were used in daily economic life (ibid., p. 31). The clear influence of writing and a notational system upon the development and organization of mathematical skills is easily discernible from these tables. Economics proved to be a motivating factor for both writing and mathematics, which mutually reinforced one another's development.

The results were tables of multiplication, reciprocals, squares, square roots, cubes, cube roots, sums of squares and cubes needed for solutions to algebraic equations and exponential functions (ibid., pp. 33–34). The sexagesimal number system 60 was developed in response to the Babylonians' concern for astronomy. The parallel between the approximately 360-day year and the 360-degree circle are obvious.

Tables of quadratic and cubic functions were prepared for civil-engineering projects of dam building, canal dredging, and the construction of attack ramps to breach the ramparts of besieged walled cities. Certain Babylonian mathematical tablets indicate that astronomy, banking, engineering, and mathematics were practiced in a systematic and scientific manner. Two types of tablets were prepared. In one set, only problems are given, but each tablet contains problems related to the other and carefully arranged beginning with the simplest cases. The second set of tablets contains both problems and their solutions worked out step by step (ibid., p. 43). The achievement of Babylonian mathematics, which has been likened to that of the Renaissance (ibid., pp. 30 & 48), is all the more remarkable when one considers the short period in which it developed and flowered: all within two hundred years or so of the major reforms in the writing system.

The existence of these tablets illustrates two important impacts of writing on science. The first is the impulse to organize information in an orderly and systematic manner. The ordering of individual words that the use of syllabic signs creates in the thought patterns of their users inspires a similar ordering of the contents of their writings. That this was critical for the development of science is beautifully illustrated by the Babylonian mathematical texts created as aids to various scientific and engineering activities.

The second impact of writing is the ability to preserve the accomplishments of one age so that they can form the basis of a later development. Little if no progress was made in Babylonian mathematics from the time of the Hammurabic explosion of knowledge to the Assyrian empire of 700 B.C. Yet the tablets preserved the knowledge that an earlier age had created and they served as the foundation for the Assyrian development.

The mathematical and scientific achievements of the Mesopotamian civilization we have just reviewed are certainly worthy of our respect and admiration. We must be careful, however, not to jump to the conclusion that this culture had solidly embarked upon the road of scientific thinking because of the progress in astronomy, mathematics, and engineering that has been described. The reader must bear in mind that the very same practitioners of this rudimentary form of science were also engaged in astrology, the reading of animal entrails, the interpretation of omens, and other forms of superstition. The early forms of science as practiced in Babylon are not a scaled down or less advanced version of science as we know it today but rather a mixture of logic, superstition, myth, tradition, confusion, error, and common sense. No distinction was made between "religious" and "scientific" thinking. "Medicine grew out of magic, and in many cases was indistinguishable from it (Cottrell 1965, pp. 169–71)." What is important about Babylonian science from a historical point of view was its influence on future generations, on the Hebrews, on the Greeks, on the Arabs, and eventually on Renaissance Europe.

The Babylonians made use of a logical mode of thought complete with abstract notions and elements of classification (Albright 1957, p. 198). Their approach was wholly empirical, however, unlike the theoretical and more analytic style of Greek science, which, according to Kramer (1959, pp. 35–36), required "the influence of the first fully phonetic alphabet." For example, the Sumerians compiled grammatical lists and were aware of grammatical classifications, yet they never formulated any explicit rules of grammar. In the field of science, lists were also compiled but no principles or laws were ever enunciated. In the field of law, a legal code was developed but never a theory of jurisprudence.

**Egyptian Writing and Science**

Like the Mesopotamians the ancient Egyptians also had a writing system and a science tradition. They also engaged in mathematics but unlike the Mesopotamians who were great at algebra the mathematical strength of Egypt was in geometry. Their writing system was not phonetic but pictographic and hence might explain why they did not achieve the same level of abstraction in algebra, which involves the manipulation of a small number of symbols.

The flooding of Nile River was extremely important to the existence of Egyptian agriculture because it supplied the water necessary for farming in a land that was otherwise a desert. The flooding also gave rise to Egyptian geometry in a round about way because of the need to measure the area of land in the possession of a landowner before the inundation of the Nile washed away all the boundary lines between properties. Rather than restore the boundary lines that were destroyed by the flooding, each landowner was provided with a new plot of land more or less in the same location as before and with a total area exactly equal to the amount of land in his possession before the flood. Because of this need to measure the area of land accurately, an empirical science arose called geometry, which literally means earth (geo) measuring (metry). Egyptian geometry is not derived from a set of axioms. There are no theorems or proofs or propositions. There are merely a set of rules that are used strictly for practical applications such as land measuring and construction calculations. They made use of the Pythagorean theorem thousands of years before Pythagoras ever proved the theorem. They did not need a proof. As long as it worked and allowed them to measure land areas accurately and carry out their engineering projects, they were satisfied. It was the Greeks who took the empirical results of Egyptian geometry and turned geometry into a set of axioms and theorems made famous by Euclid's Elements.

In addition to their abilities at geometry the Egyptians were also excellent astronomers, the knowledge of which served their agricultural needs. Agriculture also led to a number of other science based technologies such as irrigation canals and hand powered pumps, the use of yeast to make bread that would rise; pottery; glass making using soda-lime, lead, and various chemical to make tinted glass; weaving, and dyeing in which a number of chemicals were used to achieve a wide spectrum of colours. In addition to agricultural based technologies the Egyptians excelled at the metallurgy of copper, gold, silver, lead, tin, bronze, cobalt (for colouring) and iron. They also made a variety of different coloured pigments for painting. In addition to all of the chemical skills they developed must be added their ability to mummify the dead.

The Egyptians also developed incredible engineering abilities in building the pyramids, the sphinx at Giza, temples with gigantic columns, and obelisks. These engineering feats required a practical knowledge of many of the principles of physics but as with their geometry and chemistry their scientific knowledge grew out of the practical things that they did. There was not much effort made to systematize their knowledge to create a rudimentary form of science as the Greeks eventually did.

## Chinese Science

What makes the lack of theoretical science in China so puzzling is the high level of technological progress achieved there, which exceeded that of the Mesopotamians and the Egyptians that we just reviewed and the ancient Greeks who we will study in the next chapter. The list of significant scientific and technological advances made by the Chinese long before their development in the West includes the equine harness, iron and steel metallurgy, gunpowder, paper, the drive belt, the chain drive, the standard method of converting rotary to rectilinear motion, and the segmental arch bridge (Needham 1979). To this must be added irrigation systems, ink, printing, movable type, metal-barrel cannons, rockets, porcelain, silk, magnetism, the magnetic compass, stirrups, the wheelbarrow, Cardan suspension, deep drilling, the Pascal triangle, pound-locks on canals, fore-and-aft sailing, watertight compartments, the sternpost rudder, the paddle-wheel boat, quantitative cartography, immunization techniques (variolation), astronomical observations of novae and supernovae, seismographs, acoustics, and the systematic exploration of the chemical and pharmaceutical properties of a great variety of substances.

Joseph Needham carefully documented through years of historical research the contribution of Chinese science and its influence on the West. Although he championed Chinese technology he nevertheless posed the following question: "Why, then, did modern science, as opposed to ancient and medieval science, develop only in the Western world? (ibid., p. 11)" What Needham meant by "modern science," was abstract theoretical science based on experimentation and empirical observation, which began in Europe during the Renaissance.

Abstract theoretical science is a particular outgrowth of Western culture that is not more than four hundred years old. Nonabstract practical science as it occurs in ancient China, Mesopotamia, Egypt and the remainder of the world is a universal activity that has been pursued by all cultures, literate and non-literate, as part of their strategy for survival. Claude Lévi-Strauss (1960) in *The Savage Mind* gives numerous examples of elaborate classification schemes of preliterate cultures, based on their empirical observations and demonstrating their rudimentary concrete scientific thinking.

China created the most sophisticated form of technology and nonabstract science that the world knew before the science revolution in Europe during the Renaissance. Technological sophistication by itself, however, does not guarantee the development of abstract theoretical science. Other factors (social, economic, and cultural), obviously present in the West and not the East, must have played a crucial role as well. In fact in the next chapter we will show that the critical difference was the difference of the Western writing systems based on the phonetic alphabet of 20 to 30 characters as opposed to the Chinese writing system that contains thousand of characters and makes use of pictorial elements and a limited amount of phonetics.

Before delving into the impact of the Chinese writing system, let us first review the fundamental elements of Chinese science. According to classical Chinese scientific thought the universe consists of five elements: earth, water, fire, metal, and wood. The five elements are ruled by the two fundamental universal and complementary forces of yin and yang, which represent, respectively, the following pairs of opposites: cold and warm; female and male; contraction and expansion; collection and dispersion; negative and positive. The five elements and the two forces of yin and yang form a blend of opposites in which a unity emerges more through harmony than through the fiat of preordained laws (Needham 1956). Chinese scientific thought always had a mystical and mysterious aspect to it. The Confucians and Logicians, who were rational, had little interest in nature. The Taoists, on the other hand, who were interested in nature were mystics who mistrusted reason and logic. Chinese science was colored by the Taoist attitude toward nature, which is summarized by the following passage from the Huoi Nan Tzu book: "The Tao of Heaven operates mysteriously and secretly; it has no fixed slope; it follows no definite rules; it is so great that you can never come to the end of it; it is so deep that you can never fathom it (ibid., p.16)."

It is not difficult to understand how the Taoist mystical attitude toward nature might preclude the development of abstract science. We are still left, however, with the question of why those who were rational, such as the Confucists and the Logicians, were not interested in nature and why those who were interested in nature, such as the Taoists, were mystical. In other words, why wasn't there a group in China that was both rational and interested in science and nature? Eberhard (1957) offers an explanation: Science had only one function, namely, to serve the government and not its own curiosity. All innovations were looked upon as acts of defiance and revolution. The difficulty with the explanation provided by Eberhard is that it applies to the West as well. Western scientists faced the same problems in Europe. The work of Copernicus was openly contested and then suppressed by the Church, yet the Copernican revolution succeeded.

Yu-Lan Fung (1922) explains the lack of interest in theoretical science in the following terms: "Chinese philosophers loved the certainty of perception, not that of conception, and therefore, they would not and did not translate their concrete vision into the form of science." The aim of Chinese culture was to live in harmony with nature with no need to subdue it or have power over it as is the case in the West. The philosophical disposition of the Chinese was to focus on their internal reflective state rather than take the external active stance that the West adopted to develop scientific thinking. Fung's explanation is similar to that of Latourette (1964), who claimed that Chinese thinkers, unlike their Western counterparts, were more interested in controlling their minds than nature itself, whereas in the West, the opposite was true. We will see in the next chapter that the difference in the Western and Eastern writing systems also played a role.

## Chapter 4

# Physics of the Ancient Greek Era

In the last two chapters we attempted to show how the roots of scientific thinking first arose in pre-literate societies and then in Neolithic civilizations. It is with the ancient Greek thinkers, however, that the study of physics first became defined. The word physics itself is derived from the Greek word, φυσισ (phusis) meaning nature. The Greeks gave more than a name to the study of physics for it is with them that the abstract development of physics began. They are the first to apply deductive thinking to physics, to investigate the relation of physics to mathematics, and to search for a universal explanation of nature's mysteries. Although certain Greek thinkers understood the value of the empirical approach the Greeks had difficulties combining this aspect of physics with the abstract deductive theoretical aspect of physics they so highly prized. It is for this reason most likely that Greek physics did not come to full flower. However, it served as the basis for the final flowering of physics that finally occurred in Western Europe during and just after the Renaissance. Like the Renaissance thinkers there is much we can learn from the Greeks by studying both their successes and the reasons for their failures.

Greek science did not begin in isolation in fact quite the opposite is the case. Being a trading people the Greeks were in contact with the intellectual influences of other cultures such as Mesopotamian astronomy and mathematics, Hebrew philosophy and Egyptian astronomy, medicine, chemistry and mathematics. Perhaps the most important influence of all was the Egyptian discovery of geometry from their practice of land measurement. Note the word geometry comes from the Greek words for earth, geo and measure, meter. Because of the overflow of the Nile each year, which destroyed the boundaries of each persons land it became necessary to develop methods of land measurement, which led empirically to many of the results of geometry.

It was the role of the Greeks to formalize these results and derive them deductively. It is the physicist Thales, considered one of the world's Seven Wise Men by the Greeks, who first began this process of deriving the empirically based results of the Egyptians deductively. In the deductive process one makes a set of assumptions or axioms, which one considers to be self-evident truths. For example in geometry, it is assumed that the shortest distance between two points is a straight line. From these self-evident truths or axioms one then derives results using the laws of logic. An example of such a law is if $a = b$ and $b = c$ then $a = c$. The Greeks and Thales were the very first to use this method of obtaining or organizing knowledge. Up until this time knowledge was arrived at inductively i.e. by example or observation. For example, if I notice every time I put a seed into the ground a plant grows I learn by induction that seeds give rise to plants. If I notice that seeds from oranges always give rise to orange trees and seeds from lemons give rise to lemon trees I would conclude by induction that apple seeds give rise to apple trees and peach seeds to peach trees. The process of induction also involves logic. It differs from deduction, however, in that its results are not based on a set of axioms. Although the Greeks used both methods of reasoning they had a definite preference for the deductive method.

The Greek tradition of deductive geometry begun by Thales was continued by the mystic Pythagoras and his followers, who formed a brotherhood to practice the religious teachings of their master. Perhaps the best known result of their work is the Pythagorean theorem, which relates the sides of the right triangle in the accompanying Fig. 4.1 by $a^2 + b^2 = c^2$, where c is the length of the hypotenuse and a and b are the lengths of the other two sides. Perhaps the most significant discovery that Pythagoras made, however, was the relationship between harmony and numbers. He first discovered the relation between the length of a string to the frequency of the sound it emitted. He then discovered that those intervals of the musical scale that produced the fairest harmony were simply related by the ratio of whole numbers. This result led to a mystical belief in the power of numbers, as is expressed by the fragments of the Pythagorean disciple, Philolaus, who wrote:

> In truth everything that can be known has a Number, for it is impossible to grasp anything with the mind or to recognize it without.

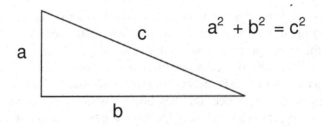

Fig. 4.1. Pythagorean Theorem

They believed that numbers were the basic stuff of the universe and that harmony controlled both the physical universe and the human soul.

Their belief in the power of numbers has been realized to the extent that almost all the phenomena described by physics is expressed in terms of mathematical equations. Their mystical approach to numbers would not have uncovered these laws, however. More than a consideration of numbers was required. Our knowledge of physics was not just arrived at using logic; observation of and experimentation in the physical world were also required. The Pythagorean infatuation with numbers can be easily understood however, when one realizes that with their discoveries regarding harmony, they were the very first to find such a dramatic connection between numbers and nature.

Despite their strong theoretical bias as illustrated by the Pythagoreans belief in numbers the Greeks also had a strong empirical tradition. This tradition was begun by Thales the very man who introduced deductive geometry into Greek thought. Thales and his Ionian followers are considered to be the world's very first physicists and philosophers. Their careful observations of nature led them to a number of conclusions held today. They believed that change and movement are caused by physical forces, that it is possible to have a void, that ice, water and steam are three different phases of water and that the changes from one phase to another are caused by condensation and rarefaction. They believed in the conservation of matter, which proved to be such an important assumption for the development of chemistry and was only recently shown not to be the case for the nuclear reactions of fission and fusion in which mass is destroyed and converted into energy.

Anaxagoras on the basis of his investigation of a meteorite concluded that the heavenly bodies are composed of rocks. This idea was extremely revolutionary because it was the common belief that the heavenly bodies were gods and goddesses. Anaxagoras was banished for his idea and almost lost his life on its account. His idea was also rejected by Aristotle, as were the concept of a void and the concept of human evolution. Anaximander argued the case for human evolution almost 2500 years before Darwin on the basis of his observations of the human embryo and on the grounds that since the newborn human child is completely helpless humans must have descended from a more primitive form of life whose young are self-sufficient. Another biological conclusion reached by the Ionian philosophers was their belief in the inherent healing power of nature.

Of all these important discoveries perhaps the most important contribution the Ionian physicists made to our intellectual heritage was their idea that all matter of the universe was composed of a single substance. Modern physics has shown that this is not the case, as there are a number of elementary particles or quarks out of which the elementary particles are composed. Still the concept of a universal such as a primary substance was the precious gift of Greek thought. We shall return to the question of why the concept of a universal arose among the Greeks. For the moment let us review their various systems. Thales, the originator of the concept, believed all matter was composed of water. His disciple Anaximander believed that all things were composed of a neutral substance he called apieron whose literal Greek meaning is infinite or limitless. He believed that the various opposite such as hot and cold, dry and wet, light and dark were all contained within this neutral substance apieron and became separated to form the various substances observed in nature. For Anaximenes the primary substance is air and the diversity of matter is explained in terms of the varying densities of air. For example, as air becomes thinner it becomes like fire and as it becomes denser it turns from air to mist to dew to water to mud to earth to stone and so on.

What we know of the philosophical systems of Thales, Anaximander, and Anaximenes does not come directly from their writings but rather from the comments about their work by later authors such as Plato and Aristotle. The first Ionian physicist-philosopher to have left a substantial fragment of his writing behind is Heraclitus whose fascinating aphorisms are still of philosophical, scientific, and literary interest.

His faith in the empirical method as well as his understanding of it are attested to in the following aphorisms:

> The things of which there can be sight, hearing and learning —
> these are what I especially prized. Eyes are more accurate

witnesses than ears. Eyes and ears are bad witnesses to men having barbarian souls. Men who love wisdom should acquaint themselves with a great many particulars. Seekers after gold dig up much earth and find little.

For Heraclitus the primary material of the universe is fire but his emphasis was more on fire as a process rather than as a substance. In reading his fragments if one thinks of fire as representative of energy then the overlap of Heraclitus' thinking and our own ideas regarding conservation of energy is amazing. He writes:

This universe, which is the same for all, has not been made by any god or man, but it always has been, is, and will be — an ever kindling fire, kindling itself by regular measures and going out by regular measures. There is exchange of all things for fire and of fire for all things, as there is of wares for gold and gold for wares.

The ideas contained in these two aphorisms almost completely parallel my personal belief that from conservation of energy and the equivalence of mass and energy generally expressed as $E = mc^2$, one is forced to conclude that the universe always was and always will be since the creation or destruction of the universe would involve the most colossal violation of energy conservation imaginable. I have underlined my belief because there are many physicists who believe in the big bang theory that the universe began some 15 billions years ago, but once again we are getting ahead of our story. More of this later.

The idea that the universe is represented by fire in the sense of process rather than substance is substantiated by the following fragments:

The phases of fire are craving and satiety. It throws apart, then brings together again; it advances and retires. Everything follows and nothing abides; everything gives way and nothing stays fixed. You cannot step twice into the same river, for other waters and yet others go ever flowing on. Cool things become warm, the warm grows cool, the moist dries and the parched becomes moist. It is in changing that things find repose. It should be understood that war is the common condition, that strife is justice, and that all things come to pass through the compulsion of strife. Homer was wrong in saying "would that strife might perish from amongst gods and men". For if that were to occur then all things would cease to exist.

The Heraclitean picture of the universe in which all things are in a state of flux is rather stark and not very comforting, even though these processes are according to him ruled by logos or reason, viz., "All things come to pass in accordance with this logos".

It is not surprising that a reaction to the Heraclitean point of view developed since as stated earlier one of the aims of a worldview is to relieve one's anxieties about the uncertainties of the world not reinforce it. Using for, perhaps, the first time in our intellectual history an internally self-consistent logical argument Parmenides developed a counterview to Heraclitean flux in which the universe was an unchanging static unit. Parmenides argued that the concept of Non-being is logically self-contradictory. "Non-being cannot be and therefore no change is allowed for if a change from the state A to state B occurs then state A will not-be. But not-being is impossible and therefore nothing changes." If this is true then concludes Parmenides what appears to our senses as change must be deception and should not be trusted. With this very simple argument Parmenides caused every subsequent Greek thinker to question the value of the empirical approach. Parmenides views were challenged by contemporaries who argued that his system did not allow motion. His disciple Zeno countered their challenge with the following proof of the non-existence of motion.

> If anything is moving, it must be moving either in the place in which it is or in a place in which it is not. However, it cannot move in the place in which it is (for the place in which it is at any moment is of the same size as itself and hence allows it no room to move in) and it cannot move in the place in which it is not. Therefore movement is impossible.

Either because of their faith in empiricism was weak or because they were so impressed by their new discovery of deductive logic, the influence of these arguments of Parmenides and Zeno on Greek thought was enormous. Instead of concluding that there are limitations to deductive reasoning as is now recognized, the Greeks, by in large, concluded that sense data could not be trusted. Anaxagoras writes "because of the weakness of our senses we are not able to judge the truth". Even the subsequent empiricists such as Empedocles and the atomists, Democritus and Leucippus pay their respects to Parmenides argument by incorporating into their world system certain immutable elements of which the universe is composed.

For Empedocles the four immutable substances of which the universe is composed are Earth, Water, Air and Fire. Since most entities are mixtures of these elements the cause of the observed change of the world is due to the fact that the four basic elements are attempting to coalesce into pure states.

The atomists Leucippus and Democritus believed in the existence of tiny particles that they called atoms, invisible to the human, which possess only the properties of size, shape and motion. Each object is composed of a different combination of atoms. Although individual atoms cannot change their properties the objects of which they are composed can change as different combinations of these atoms form. The prediction of atoms some 2500 years before their actual discovery is a tribute to the richness of the thought of the early Greek physicists. The fact that our modern days atoms are mutable is not a shortcoming of our ancient predecessors but rather a failure on our part in labeling. We should have reserved the name atom for elementary particles such as proton, neutrons and electrons and called the objects that we now label as atoms by some other name. Or perhaps we should have reserved the word atom for quarks because as we will discover neutrons decay into protons, electrons and neutrinos, but more of that later.

The effect of Parmenides paradox on other thinkers was more devastating and contributed in my opinion to the demise of Greek physics. Although the Greek empirical tradition continued its value was seriously called into question. For example, Plato's mistrust of the senses was so great that he schizophrenically created two worlds, one the world of sense perception, which he wants us not to trust and the other, the world of ideas or forms, the only world where truth and knowledge is possible.

The physics of Aristotle represents a synthesis of Ionian materialism and the Platonic concern for form for he held that form and matter are inseparable. It is important to review his physics not because he represents the highest achievement of Greek science (in certain ways his work is retrograde) but because of his enormous historical influence particularly with the thinkers of the Middle Ages with whom modern science began. His influence, in part, was due to the methodical way in which he argued away any other point of view but his own, such as the atomists' concept of a void. The other reason is, like Plato, he founded a university, the Lyceum, which propagated his point of view.

The Parmenidean influence on Aristotle expresses itself, as with Plato, in a dichotomy. He divides the universe into two concentric spherical regions. At the center is the imperfect Earth constantly in change surrounded by the sphere of perfection, the immutable heavens. In the sub-lunar region of imperfection, matter is composed of the four Empedoclean elements, which are trying to arrange themselves into four concentric spheres in which the element earth is at the very center followed by water, air, and fire somewhat like the way these substances are arranged into the solid earth, the oceans and the atmosphere with fire having the property that it rises to the top. This model explains gravity as the tendency of earthy things to gather together. Change in general is explained in terms of the propensity of the Empedoclean elements to coalesce.

Aristotle postulated a fifth element, the eternal and unchanging aether, which completely fills the heavens so there is no empty space. "Nature abhors a vacuum." The heavenly bodies move naturally without the assistance of any force in perfectly circular orbits, the circle being the most perfect shape imaginable. Aristotle astronomy was adopted from the work of Eudoxus. It was later developed by Ptolomey who changed details of the system but left the basic structure unchanged.

Motion on Earth unlike the heavenly motion tends to be rectilinear and constantly requires the action of a force. Since a concept of inertia is missing not even constant rectilinear motion can be explained without the action of some force. For example, Aristotle explains that the reason an arrow continues to move once it has lost contact with the bow is that as the arrow moves it creates a void, which nature abhors and hence the arrow is pushed along by the air rushing in to fill up the vacuum. If a force produces a constant speed then we are left with the puzzle of why falling objects accelerate. Aristotle claims that since the falling object is traveling back to its proper place in the universe the joy of its returning home makes it speed up.

With the exception of the atomists and the astronomers, Heracleides (not to be confused with Heraclitus) and Aristachus, to be discussed below the worldview of Aristotle held swayed until the Copernican revolution of the 16th century. Some thinkers have considered this a tribute to Aristotle who they say was 2000 years ahead of his time. From my point of view, however, it is only a reflection of his ability to suppress ideas other than his own by logically arguing them away. He thereby created an atmosphere that was not inductive to new ideas as is illustrated by the reception of the ideas of the two post-Aristotelian astronomers Heracleides and Aristachus. Heracleides proposed that the daily rotation of the heavens could be more easily explained in terms of the rotation of the Earth rather than the entire heavens. He also proposed that the planets Mercury and Venus orbit the sun instead of the Earth. Aristachus incorporating these ideas, also proposed that the Earth orbited the sun and not vice-versa. His contemporaries found it difficult to accept the movement of the Earth. Also the enormous distance to stars that his system implied was difficult for them to comprehend. Some 2000 years later, however, the ideas of Aristachus formed the foundation of the Copernican system.

While no major new worldviews developed after Aristotle, the Greeks achieved a number of solid results in which mathematical concepts played an important role. These included a formulation of the mathematical laws of simple machines such as the lever, the wedge, the screw and the pulley, begun by Archimedes and completed by Hero, as well as Archimedes' advances regarding hydrostatics including his explanation of buoyancy.

The Roman interest in physics was almost exclusively in terms of practical applications. Their engineering achievements such as their aqueduct system supplying Rome with millions of gallons of fresh water, their sewage system, their road system and their harbors are all worthy of mention. Little can be said about the scientific achievements during the early Christian era, which immediately followed the Roman period. Interest in science declined to an even greater extent, as theology became the dominant concern of the day. The Greek scientific tradition continued in the East, however, by Arabic scholars whose major contribution was the development of algebra and chemistry. To them we owe thanks for the transmission of the concept of zero, a non-trivial concept invented by Hindu mathematicians. They also preserved much Greek learning that might have been lost otherwise.

When the resurrection of interest in science took place in Europe during the Renaissance the three sources of ancient Greek learning were from those original Greek works that survived and from the comments and translations of both Latin and Arabic scholars. Before turning to the rise of modern science in Europe let us first examine the question of why abstract scientific thinking first began in Ancient Greece.

In the last chapter we pointed out that the scientific achievements of the ancient cultures we studied were intimately connected with their technological activities. One is tempted like a number of authors to conclude that the Greek achievement of an abstract science is connected with their knowledge of the technical achievements of the people they were in contact with such as the Egyptians and the Mesopotamians. If this is true one wonders why the Chinese did not develop abstract physics since their technological superiority to the Greeks is attested to by their invention of clocks, iron casting, paper, block printing, movable print, silk, animal harnesses, irrigation canals, suspension bridges, gun powder, guns and porcelain to mention a few. Joseph Needham in his book The Grand Titration claims the Chinese played an important role in the developments of science in the West as information of their discoveries reached Europe. I am sure to some extent this is true but one cannot help but ask the question if technology plays such an important role why didn't the Chinese develop science themselves.

When I first formulated this question when I wrote the first draft of this book way back in 1973 my answer to this question was in terms of two influences, which were felt much more strongly by the Greeks than the Chinese, namely codified law and monotheism. The first of these influences, the codification of law in the West, began in Babylonian under Hammurabi. In China behavior was guided more by tradition, moral persuasion and social pressure than by a legal code. The Chinese had laws but they were not codified. The law was an important part of Greek life with a philosopher often playing the role of the lawgiver in his society. It is not much of a stretch that the concept of human law would lead one to develop the notion of the laws of nature. The analogy between these two concepts of law was expressed by a number of the early Greek physicists. Anaximander wrote: "The Unlimited (Apieron) is the first principle of things that are. It is that from which the coming-to-be takes place, and it is that into which they return when they perish, by moral necessity, giving satisfaction to one another and making reparation for their injustice, according to the order of time."

The other influence that I identified was also felt much more strongly in the West than the East and that was the Hebrew concept of monotheism. Before the Jewish concept of God people believed that a god was localized in a place such as on the top of a mountain or under the sea. The influence of the god and hence the laws pertaining to his worship were only local in nature. They held sway over the small territory in which they were worshipped. With the concept of an omnipresent God whose law applied everywhere the idea of a universal law developed. All of the early Ionian physicist and later empiricists were monotheistic and believed in a universal law as is illustrated by the following quotes from Heraclitus, "All things come to pass in accordance with this Logos" and from Anaxagoras, "And Mind set in order all that was to be, all that ever was but no longer is, and all that is now or ever will be."

These fragments illustrate the intimate connection for the Greek physicists between the belief in universal law and monotheism, which parallels the Jewish relationship of Jehovah and the Law. The Greek deity is usually an abstraction of reason as is illustrated by the way Heraclitus and Anaxagoras refer to him as Logos and Mind respectively.

This was my explanation for why abstract science began in the West as of 1974 when I first met my colleague Marshall McLuhan, an English professor at the University of Toronto and the communications theorists who was famous for his one-liners: "the medium is the message", "the global village" and "the user is the content." McLuhan (1962 & 1964) together with Harold Innis (1971 & 1972), whose works McLuhan built upon, were the founders of a tradition known as the Toronto School of Communications. The Toronto School established at the University of Toronto in the fifties explored the ways in which media of communication, including the alphabet, have shaped and influenced human culture and its various social institutions. In particular, McLuhan showed that the use of the phonetic alphabet and the coding it encouraged led the Greeks to deductive logic and abstract theoretical science. The tradition that began as the Toronto School of Communication now has a much broader geographic base and has given rise to the term media ecology (see www.media-ecology.org).

In 1974 McLuhan, having heard of my Poetry of Physics course, invited me to lunch at St. Michael's College at the University of Toronto. He asked me what I had learned from my Poetry of Physics project. I told him of my attempt to explain why abstract science began in the West instead of China because of the traditions of codified law and monotheism as I described above. McLuhan agreed with me but pointed out that I had failed to take into account the phonetic alphabet, another feature of Western culture not found in China, which had also contributed to the development of Western science. Realizing that our independent explanations complemented and reinforced each other, we combined them in a paper entitled "Alphabet, Mother of Invention" (McLuhan and Logan 1977) to develop the following hypothesis:

> Western thought patterns are highly abstract, compared with Eastern. There developed in the West, and only in the West, a group of innovations that constitute the basis of Western thought. These include (in addition to the alphabet) codified law, monotheism, abstract theoretical science, formal logic, and individualism. All of these innovations, including the alphabet, arose within the very narrow geographic zone between the Tigris-Euphrates river system and the Aegean Sea, and within the very narrow time frame between 2000 B.C. and 500 B.C. We do not consider this to be an accident. While not suggesting a direct causal connection between the alphabet and the other innovations, we would claim, however, that the phonetic alphabet (or phonetic syllabaries) played a particularly dynamic role within this constellation of events and provided the ground or framework for the mutual development of these innovations.

The effects of the alphabet and the abstract, logical, systematic thought that it encouraged explain why abstract science began in the West and not the East, despite the much greater technological sophistication of the Chinese, the inventors of metallurgy, irrigation systems, animal harnesses, paper, ink, printing, movable type, gunpowder, rockets, porcelain, and silk.

There is a reason why the alphabet has had such a huge effect on Western thinking. Of all the writing systems, the phonetic alphabet permits the most economical transcription of speech into a written code. The phonetic alphabet introduced a double level of abstraction in writing. Words are divided into the meaningless phonemic (sound) elements of which they are composed and then these meaningless phonemic elements are represented visually with equally meaningless signs, namely, the letters of the alphabet. This encourages abstraction, analysis (since each word is broken down into its basic phonemes), coding (since the sounds of spoken words are coded by visual signs), and decoding (since those visual signs are transformed back to spoken sounds through reading).

The twenty-six letters of the English (or Roman) alphabet are the keys not only to reading and writing but also to a whole philosophy of organizing information. We use the letters of the alphabet to order the words in our dictionaries, the articles in our encyclopedias, the books in our libraries, and the files on our computers. These systematic approaches to coordinating information based on the medium of the alphabet have suggested other forms of classification and codification that are part and parcel of Western science, law, engineering, economics, and social organization.

Thus we see there is more to using the alphabet than just learning how to read and write. Using the alphabet also entails the ability to: 1) analyze, 2) code and decode, 3) convert auditory signals or sounds into visual signs, 4) think deductively, 5) classify information, and 6) order words through the process of alphabetization. Each of these skill sets was essential to the development of abstract science. These skills are the hidden lessons of the alphabet that are not contained (or at least not contained to the same degree) in learning the Chinese writing system or any of the other non-alphabetic writing systems.

# Chapter 5

# The Roots of the Scientific Revolution

Although the science revolution took place in Renaissance Europe the roots of this revolution can be said to have also taken place in Ancient India and the Medieval Islamic world. We turn first to the contribution of Hindu mathematicians and then examine the contributions of Islamic science.

## Hindu and Buddhist Mathematics, the Invention of Zero and the Place Number System

Hindu and Buddhist mathematicians invented zero more than 2,000 years ago. Their discovery led them to positional numbers, simpler arithmetic calculations, negative numbers, algebra with a symbolic notation, as well as the notions of infinitesimals, infinity, fractions, and irrational numbers all of which were essential elements in the breakthroughs of Copernicus, Kepler, Galileo and Newton.

The historians of mathematics have always been puzzled that the germinal idea of zero was a discovery of the Hindus and not the Greeks. The explanation of this fact does not lie in an examination of Greek mathematics but rather in a comparison of Greek and Hindu philosophy. Paradoxically, it was the rational and logical thought patterns of the Greeks that hindered their development of algebra and the invention of zero (Logan 2004). Let us recall that Parmenides using logic argued that non-being could not be and that Aristotle also argued that a vacuum could not be. These philosophical notions I would claim created an environment that discouraged the conceptualization of zero. Non-being was a state that Hindus and Buddhists actively sought, on the other hand, in their attempt to achieve Nirvana, or oneness with the whole cosmos

37

and they, unlike the Greeks, did not have an intellectual tradition of formal logic. They were less constrained in their thinking and more imaginative, which proved to be an invaluable asset and led to the development of zero.

The development of place numeration, whereby all numbers can be represented by the ten symbols 0, 1, 2, 3, 4, 5, 6, 7, 8, and 9 is probably the most important application that was made of the zero or sunya symbol. Our present number system was invented by the Hindus and transmitted to Europe by Arab and Persian scholars.

Hindus hit upon zero when they were notating the results of their abacus calculations. When they arrived at a result like 602, on their abacus, they looked at the beads and saw that there were 6 hundreds and 2 ones and instead of writing out 602 using the symbol for 6 followed by the symbol for 100 followed by the symbol for 2 they wrote 6 sunya 2, where sunya means literally "leave a space" in Sanskrit so that 6 sunya 2 was read as 6 hundreds, no tens and two ones. They then notated sunya with a dot and then later with a circle, 0. This notation evolved into the symbol for zero and the place number system, which greatly simplified arithmetic operations like addition, subtraction, multiplication and division. By placing the sunya (zero) sign over a number the Hindu mathematicians developed a notation for negative numbers so that −7 was read as 7 below zero. The sunya sign was also used to denote the unknown and was used to develop algebra, i.e. equations involving an unknown. Sunya was also used to talk about infinitesimals and infinity, which was achieved by dividing any numeral by sunya.

The idea of sunya and place numbers was transmitted to the Arabs who translated sunya or "leave a space" into their language as sifr. The mathematicians of Baghdad adopted the Hindu system around A. D. 1000. These ideas first arrived in Europe through the Italians who traded with the Arabs and adopted their numerical notation system. This is how we came to regard our number system as Arabic numerals. The term cipher came directly from the Arabic sifr. The term cipher means both zero and a secret code, the latter denotation because the use of Arabic numerals was at first forbidden by the Roman Catholic Church but was used by the Italian merchants as a secret code. In order to distinguish between the entire number system, which was called cipher and its unique element, 0, the term zero, short for zepharino (the Latinized form of cipher), came into use to denote 0.

Many of the Hindu techniques for complicated calculations like square and cube rooting that were developed using the place number system were transmitted to Europe by the Arab mathematician, al-Khwarizmi, from whose name is derived the term algorithm, meaning a procedure for performing a mathematical calculation. The scientific revolution of the Renaissance could never have taken place if these simpler modes of calculation had not been made possible by the Hindu place number system.

The scientific revolution also benefited from the Hindu development of algebra. The Hindus' successes in developing algebra, like their success with zero, stemmed from their ability to work intuitively without being unnecessarily held back by the need for logical rigor. The essential element in their development of algebra was their invention of zero, which proved to be a powerful mathematical concept.

## Islamic Medieval Science

After the fall of Rome learning in Europe with the exception of theology went into a decline. However, in the Islamic world in which Arabic was the lingua franca a very vital level of scientific activity took place between the 7th and 16th centuries. "The only effective link between the old and the new science is afforded by the Arabs. The dark ages come as an utter gap in the scientific history of Europe, and for more than a thousand years there was not a scientific man of note except in Arabia (Lodge 2003)."

Baghdad, the capital of the Abbasid dynasty from 750 to 1258, became a center of learning. The philosophical and scientific works of the Hellenistic world were translated and introduced to the Muslims. This stimulated new and original research and study in which the Arabs made significant and lasting contributions (ibid., p. 65). Baghdad became, as an Arab historian described it, "the market to which the wares of the sciences and arts were brought, where wisdom was sought as a man seeks after his stray camel, and whose judgment of values was accepted by the whole world" (Gibb 1963, p. 46). Baghdad soon possessed a library and an academy that in many ways rivaled the original library at Alexandria (Hitti 1964, p. 117).

One of the factors that contributed to the intense literary activity in the Arab world was the sudden availability of paper in the mid-eighth century, replacing the more expensive media of parchment, papyrus, and leather (Gibb 1963, p. 41). The Arabs borrowed this know-how from the Chinese and eventually passed it on to the West.

The contribution of the Arabs to the overall development of modern science was twofold. First, there are the advances and discoveries they made totally on their own. And second, there is the role they played in the preservation of and transmission to Europe of the scientific accomplishments of ancient Greece, India, Persia, and China. The Arabs built upon the base of Greek learning contained in the Syriac and Persian literature both in translation and in the original Greek.

There were many significant advances in medicine, engineering, mathematics such as algebra (actually an Arabic term) and the sciences of chemistry and astronomy. They have been credited with laying the foundations for an empirical approach to science with their contributions to the formulation of the scientific method, which was exemplified by and their experimental and quantitative approach to scientific inquiry.

> According to the majority of the historians al-Haytham was the pioneer of the modern scientific method. With his book he changed the meaning of the term optics and established experiments as the norm of proof in the field. His investigations are based not on abstract theories, but on experimental evidences and his experiments were systematic and repeatable (Gorini 2003).

Islamic scientists excelled at chemistry pioneering such procedures as distillation, liquefaction, oxidization, crystallization, filtration and purification leading to products like soap, shampoos, perfumes and more importantly medicines. The term chemistry is a term that derives from the Arabic alkimiya or alchemy where the prefix al means "the". While alchemy eventually came into disrepute in Europe, historians of science recognize that alchemy as practiced by the Arabs laid the foundations for modern chemistry.

Another area where Arabic science was extremely successful and went far beyond the Greeks was medicine and pharmacology, particularly in observation, diagnosis, and treatment with drugs. They also excelled in surgery. "It was they who established the first apothecary shops, founded the earliest school of pharmacy and produced the pharmacopoeia" (Hitti 1964, p. 141). Important contributions were also made in agriculture, magnetism, geography, optics, ophthalmology, and astronomy.

They also contributed a great deal in mathematics and logic including algebra. The term algorithm is a latinization of Al-Khwarizmi the great algebraist whose book title Kitab al-Jabr gave rise to the term algebra. Arab mathematicians transmitted the Hindu place number system to Europe.

Arabic astronomical observations were more accurate than those previously made because of improvements in their astronomical instruments. They increased the size of the armillary sphere and astrolabe and thus reduced the errors of observation. They were able to calculate the radius of the earth and the meridian degree with only 1 percent error, by A.D. 820. In the meantime, Europe still slumbered under the illusion that the earth was flat. They foreshadowed the work of Copernicus introducing the notion of the daily rotation of the Earth.

They also made significant contributions in optics, geology, zoology and physics. They were excellent botanists, which contributed to their expertise in agricultural practices.

Although the Arabs made vital contributions, they fell short of the development of modern science. "They introduced the objective experiment, a decided improvement over the hazy speculations of the Greeks. Accurate in the observation of phenomena and diligent in the accumulation of facts, the Arabs nevertheless found it difficult to project proper hypotheses and draw truly scientific conclusions" (Hitti 1964, p. 147). They did, however, lay the foundations for modern science, which they transmitted to Europe.

## Medieval European Science

The development of modern science between the 13th and 17th centuries in Europe occurred in two stages. During the first stage there was a renewal of interest in ancient Greek learning. The philosophy of Aristotle once again dominated the thinking of the time. His authority, however, was partly neutralized by the Church. As a result two new theories arose based on a reinterpretation of the existing data rather than new experimental results. One was the impetus theory of motion and the other, more revolutionary in nature, was the Copernican heliocentric theory of the universe. The second phase of the scientific revolution was brought about as a result of the development of experimental technique.

Before this period physicists had based their generalizations on observations of nature. However, beginning with Galileo physicists become more aggressive in making observations as a result of the experiments they conducted. Even the astronomical observations made by Tycho Brahe, which led to the results of Kepler and Newton had never been made so carefully and systematically. Before turning to this question let us first investigate how the renewal of interest in ancient Greek learning in general and science in particular played out.

Interest in science waned in Europe with the coming of Christianity as men's interest turned to theology. Christian theology at first adopted the philosophy of Plato synthesized by the Neo-platonist Plotinus. The emphasis of this theology was on faith as is exemplified by the teaching of St. Augustine. The Islamic thinkers, on the other hand, also adopted the philosophy of Aristotle. In the 12th and 13th century Christian theologians begin involving reason to defend the faith and like the Islamic thinkers also adopted the philosophy of Aristotle. The synthesis of Christian theology and Aristotelian philosophy was begun by Roscelin and Abelard and culminated with St. Thomas Aquinas whose teachings still forms the basis of present day Catholic theology.

With its adoption by the Church, Aristotelian philosophy once again played a dominant role in the thinking of the time. But now that authority had to be shared with the Church, which had certain dogmas to defend such as the omnipotence of God. It was for this reason that the Bishop of Paris in 1277 ruled that three of Aristotle's concepts that purported to limit the powers of God were wrong. These included the notion that a void is impossible, that the universe is finite and that a plurality of worlds cannot exist. The influence of this ruling was to neutralize the absolute authority that the work of Aristotle had commanded.

Probably the first effect of this new atmosphere was the change in the 14th century of the Aristotelian concept of motion proposed by three of the proponents of the impetus theory, namely, Jean Buridan of Paris, Albert of Saxony and Nicholas of Oresme. Aristotle believed that all earthly motion required the constant application of a force. According to the impetus theory the forerunner of our modern theory of inertia, if an impetus is given to a body it will move of its own accord for some time but it eventually comes to rest once the impetus has worn off. It is similar to putting an iron sword in the fire. It is hot while it is in the fire and retains its heat for a while after it is withdrawn from the fire but eventually it cools off. In the same way, according to the new impetus theory once a body is in motion it will continue to move for a while until its motion or impetus wears off and it gradually comes to rest.

The acceleration of a falling body is explained as due to impetus continually being added to a body due to its weight. The ability of a body to take on or retain impetus was thought to be proportional to its density. Thus a feather falls more slowly than a stone but a large stone falls at the same rate as a small stone. The heavenly bodies that moved above the atmosphere did not lose impetus because of an absence of air resistance. The formulators of impetus were just on the verge of the concept of inertia, the idea that a body will remain in a state of constant rectilinear motion until some force changes its motion. The concept of impetus did not arise in response to new observations but rather from a different interpretation of the known facts.

# Chapter 6

# Mechanics, Planetary Motion and the Modern Science Revolution

## The Copernican Revolution

The Copernican heliocentric theory of the universe posed an even greater challenge to Aristotelian physics than the ideas of Jean Buridan Albert of Saxony and Nicholas of Oresme. His theory also grew out of a reinterpretation of the data rather than new observations. Copernicus (1473–1543) was not the best of observers, accepting good and bad observations indiscriminately. He was disturbed, however, by certain discrepancies between the Ptolemaic system and the existing experimental data. Although his system removed many of these discrepancies others remained. There were other motivations, however.

Copernicus was disturbed by the contradiction between Ptolemy's claim that the Earth was at the center of the universe and the actual details of the Ptolemaic system. In adhering to Aristotle's principle that the heavenly bodies move in circles, Ptolemy reduced the motion of the planets to combinations of uniform circular orbits called epicycles. An epicycle is generated by a circular orbit whose center moves about a second circle and that circle could be moving about a third circle as illustrated in Fig. 6.1 on the next page. Copernicus could accept the idea of epicycles and indeed incorporated them into his own scheme. What disturbed him about the Ptolemaic system, however, was the fact that the Earth was not actually at the center of the epicyclical orbit of the Sun and planets but slightly displaced from the center and hence Ptolemy's claim that the Earth was exactly at the center of the universe was a deception.

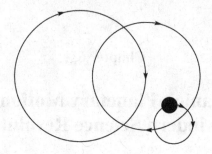

Fig. 6.1

A third motivating factor was Copernicus' mystical feelings towards the Sun, which is evident from the following passage of his book De Revolutions Orbium Coelestium in which Copernicus describes the location of the heavenly bodies in his scheme of the universe.

> The first and highest sphere is that of the fixed stars, containing itself and everything and therefore immovable, being the place of the universe to which the motion and places of all other stars are referred. Next follows the first planet Saturn, which completes its circuit in thirty years, then Jupiter with a twelve year's period, then Mars, which moves around in two years. The fourth place in the order is that of the annual revolution, in which we have said that the Earth is contained with the lunar orbit as an epicycle. In the fifth place Venus goes around in nine months, in the sixth Mercury with a period of 80 days. But in the midst of all stands the Sun. For who could in this most beautiful temple place this lamp in another or better place than that from which it can at the same time illuminate the whole? Which some not unsuitably call the light of the world, others the soul or ruler. Trismegistus calls it the visible god, the Electra of Sophocles, the all-seeing. So indeed the Sun, sitting on the royal throne, steers the revolving family of stars. (With the word star Copernicus refers to both the fixed stars and the planets. We still call the planet Venus or Mars the morning or evening star.)

Not only does Copernicus have strong mystical feelings for the Sun but also for nature in general as is exhibited by another passage from De Revolutionius. "The wisdom of nature is such that it produces nothing superfluous or useless but often produces many effects from one cause."

It is obvious from this passage that Copernicus was also motivated by a desire to achieve a simpler picture of heavenly motion. By assuming that the Earth and the other planets orbit the Sun as well assuming that the Earth rotates about its axis every 24 hours Copernicus achieved a simpler description of the heavens in which only 34 circular wheels instead of 80 were needed. Instead of the entire heavens rotating every 24 hours all of this motion was simply explained in terms of the Earth's rotation. Whereas in the Ptolemaic system the stars and the planets turned in opposite directions and the movement of the Sun and the planets was completely uncoordinated, in the Copernican system the stars are fixed; the Earth and the planets orbit in the same direction with the period of their orbit related to their distance from the Sun. In addition to its greater symmetry another advantage of the system is that is also made calculations and predictions easier.

In spite of its distinct advantages the Copernican system did not at first gain many adherents mainly because it contradicted the existing view of physics. It was difficult to image that a body as heavy as the Earth could move. It was much simpler to conceive of the motion of the heavenly bodies, which were thought to be light and ethereal as had been proposed by Aristotle. With the new system the tidy explanation of gravity was also sacrificed. Finally adoption of the Copernican view required an enormous increase of the distance to the fixed stars. If the distance to the stars of the Ptolemaic system were retained then one would be able to detect parallax, the shift in the position of the stars due to the Earth's motion. This affect went undetected until the nineteenth century.

Copernicus realized the difficulties thinkers would have in adopting his view when he wrote, "If all this is difficult and almost incomprehensible or against the opinion of many people, we shall, please God, make it clearer than the Sun, at least to those who also know mathematics."

He tried to justify the physics of his scheme by replacing Aristotelian dynamics with his own principle of circularity and sphericality. Copernicus argues that it is the natural tendency of all things to consolidate themselves so as to form a spherical shape. It therefore follows that not only does the Earth possess gravity but so does the Sun, the Moon and the other planets. It is from this concept of universality that our modern theory of gravity eventually developed. At this stage, however, gravity is strictly local; there is no gravitational interaction between the various heavenly bodies. Their circular motion as well as their rotation about their axis is regarded by Copernicus to be a result of their spherical shape. No further explanation is required.

Copernicus was unable to convince his contemporaries, however, with these arguments. He was also unable to reply satisfactorily to some of the objections they raised such as the claim that the rotation of the Earth would literally tear the Earth apart because of the centrifugal force and the claim that the rotation of the Earth from west to east would imply a continual wind from east to west. The eventual acceptance of his theory would require the downfall of Aristotelian physics that future observations and experiments finally brought about. In the intervening period, however, there was still a dedicated band of followers supporting the Copernican point of view.

In addition to scientific difficulties alluded to; the heliocentric viewpoint also gave rise to philosophical and theological difficulties. It is ironic that at the same time the Church adopted the Copernican length of the year as the basis of the Gregorian calendar reform of 1582, they were condemning the Copernican worldview because it conflicted with Scripture; for it is written that the Sun rises and the Sun sets. Even though it is now more than 350 years since the acceptance of the idea that the Earth moves and the Sun stands still, we still refer to the Sun rising and setting.

The objection of the church went much deeper than this Scriptural conflict, however. The crucial problem was the role man played in the new Copernican universe. In the old Ptolemaic system man was at the center of the universe with the heavens rotating about him for his pleasure. It was too comforting a worldview for the egocentric spirits of the Renaissance to give up without a struggle. The idea of hurtling through empty space on the space ship Earth is an idea that appeals to our age, but to the thinkers of Copernicus' time, on the other hand, it was a frightening thought.

Travel through empty space appealed to one man of this age, however, as is illustrated by the poem L'Envoi by Giordano Bruno.

> Who gives me wings and who removes my fears
> Of death and fortune? Who inflames my heart?
> Who breaks the chains and makes the portals start
> Whence but a rare one, freed at last, appears?

Time's Children and his weapons, ages, years,
Months, days, and hours, all that host whose art
Makes even adamant and iron part
Have now secured me from his fury's spears.
Wherefore I spread my wings upon the air
No crystal spheres I find nor other bar
But flying to the immense I cleave the skies
And while from my small globe I speed elsewhere
And through the ethereal ranges further rise
I leave behind what there is seen from far.

Translated by W.C. Greene

Bruno was a physicist, a poet and a mystic. He traveled extensively through Europe spreading the teaching of Copernicus. Bruno's own pantheistic mystical speculations went way beyond the Copernican point of view. Bruno believed that the universe was infinite in size and eternal. He taught that each of the stars was also a sun, a center of a solar system, which contained planets, some of which like the Earth were inhabited by intelligent creatures. These ideas, some of which are not even universally accepted today, were totally rejected by Bruno's contemporaries including the supporters of Copernicus. Bruno's imagination was more than his society could cope with. He so infuriated the Church, that upon his return to Italy he was arrested by the Inquisition and imprisoned for eight years. He steadfastly refused to recant his point of view or confess to the error of his way. In fact he bravely defended his position responding to his interrogators in the following manner:

I hold the universe to be infinite, as being the effect of infinite divine power and goodness, of which any finite world would have been unworthy. Hence I have declared infinite worlds to exist beside this our Earth, I hold with Pythagoras that the Earth is a star like all the others, which are infinite, and that all these numberless worlds are a whole in infinite space, which is the universe. Thus there is a double sort of infinity, in size of the universe and in number of worlds; this it is which has been understood to disagree indirectly with the truth according to faith.

Because of his belief in these ideas and his refusal to recant them he was bound to the stake at Campo di Fiori on February 19, 1600 and burned alive.

## Tycho Brahe's Contribution

Copernicus and Bruno, whose ideas form the basis of our modern point of view, actually represent the end of an era. They and the formulators of the impetus theory brought about great change by looking at old facts in new ways. The tremendous advances made by their successors, Tycho Brahe (1546–1610), Galileo Galilei (1564–1642), Johannes Kepler (1571–1630) and Isaac Newton (1642–1727) using new facts characterized a change in the technique of physics in which the need for careful systematic observation and experimentation was recognized. Tycho Brahe wrote, "Only through a steadily pursued course of observations would it be possible to obtain a better insight into the motions of the planets and decide which system of the world was correct."

And in criticizing those whose physics consisted mostly of speculation, he wrote, "O foolhardy astronomers, O exquisite and subtle calculators, who practice astronomy in huts and taverns, at the fireplace, in books and writings, but not in the heavens themselves. For very many do not even know the stars. And yet they would go to the stars."

Tycho Brahe went to the stars and made observation almost every night of his adult life. He probably made more observations than any other astronomer before him and perhaps since. His observations were made without the use of the telescope, which had not yet been invented. He did use instruments, however, and in fact built an observatory on the isle of Hveen in Denmark where he did most of his life's work.

Tycho Brahe recognized the advantages of the Copernican system over the Ptolemaic system. He could not accept the concept that the Earth moved, however, and a result constructed another system, which was a compromise of the two world systems. In Brahe's system the Earth is static and at the center of the universe. The Sun, the Moon, and the fixed stars all orbit the Earth. The five planets Mercury, Venus, Mars, Jupiter and Saturn on the other hand, orbit the Sun. Brahe's system served as a bridge between the two systems. His greatest contribution, however, was the wealth of experimental information that he left behind. It was his observations as we shall see later that served as the experimental basis of Kepler's great work.

## The Contribution of Galileo Galilei

The importance of Brahe's observational work cannot be under-estimated. But it is to Galileo Galilei that we owe our greatest debt of appreciation for the development of the experimental technique, which so revolutionized physics. Various influences have been suggested to explain this development by Galileo. Among these the art of alchemy is often cited. There is little doubt of its influence in the development of experimental technique particularly in the field of chemistry. It is also claimed that artists, who, like Leonardo da Vinci, were scientifically oriented, provided stimulus to the development of experimental technique. Artists used experimental methods in developing the technical aspects of their art such as the manufacture of their pigments. Perhaps more important was the development of perspective by the painters of the Renaissance. Their approach to this problem was experimental in a true fashion and, no doubt, provided a model for physics. Perhaps the greatest influence of all was the widespread interest during the Renaissance for technical and mechanical devices. Galileo, an inventor, developed a number of these devices himself. Some of these such as the geometric and military compass, the pulsilogium, a device to measure the human pulse, the magnetic compass and the pendulum regulator of a clock work were developed for practical applications, whereas others served as instruments in his experimental work such as the telescope, the microscope, the pendulum, the thermoscope (a thermometer without a scale) and the giovilabio, a device which computed distances and periods of Jupiter's satellites.

Galileo did a number of experiments investigating the nature of motion by using pendulums and inclined planes. Through his experimentation he discovered that aside from air resistance all bodies accelerate uniformly when falling to Earth. Apparently the famous story that he made this discovery by dropping objects from the Leaning Tower of Pisa is a myth. The actual discovery was made by rolling objects down an inclined plane. His investigation of the pendulum is said to have begun with his observation of a chandelier swinging back and forth in church one morning. He was fascinated by the regularity of the motion, which, he saw, could be used as the regulator of a clockwork.

Galileo's experiments and observations lead him to formulate the parabolic law of motion of a projectile in the Earth's gravitational field. These studies also lead him very nearly to a formulation of inertia, the concept that a body remains at rest or in uniform straight-line motion unless a force acts upon it. This concept, which plays such a central role in Newton's formulation of classical mechanics, was not clearly enunciated until Descartes. Another of the important contributions made by Galileo, as is exemplified by his work on projectiles, is the incorporation of mathematics into his description of nature. There is no doubt he was influenced by Euclid and Archimedes when he wrote, "trying to deal with physical problems without geometry is attempting the impossible" or "the book of Nature is written in mathematical characters."

Perhaps Galileo's most fascinating experimental work was done with his invention of the telescope. His results established the Copernican system on a firm experimental basis as he laid to rest the Aristotelian concept that the heavens are unchanging and made of the fragile substance aether. With the appearance of a new star in 1572 and the new comet of 1577, came the first empirical evidence against the concept of the immutable heavens. These two events carefully recorded by Brahe were not enough to change Brahe's attitude or that of the other scholars towards Aristotle's physics. More evidence was required and this was provided with Galileo's telescopic observations of mountains on the Moon like those found on Earth and his observation of the appearance, movement and disappearance of sunspots (which we have since discovered are electrical storms on the surface of the Sun). Finally, his discovery of the moons of Jupiter and the rings of Saturn showed that the Earth was not the only planet in the solar system with a satellite, which had been used as an argument against the Copernican scheme.

## Kepler's Three Laws of Planetary Motion

At the same time Galileo was establishing the Copernican system with his telescopic discoveries, Johannes Kepler also a supporter of Copernicus, was working on the mathematical problem of finding the actual orbits of the planets about the Sun. Kepler made use of the enormous amount of data collected by Tycho Brahe for whom he worked during the last days of the Dane's life. Instead of trying to describe the orbits as superpositions of circles or epicycles, Kepler tried to fit the orbits using different geometrical forms. After many unsuccessful attempts he finally discovered that the actual orbits of the planets are ellipses with the Sun sitting at one of the two foci of the ellipse as is shown in Fig. 6.2. This constitutes Kepler's first law of planetary motion. The eccentricities of the ellipses are very small so that they are almost circular.

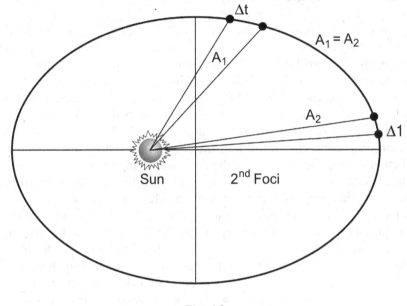

Fig. 6.2

He discovered that the same area is swept out by the planet in its orbit in equal the orbits time intervals as is shown in Fig. 6.2. This is Kepler's second law of planetary motion. He also showed that the orderly relation between the distance of planets from the Sun and the period, T, of their orbit (the time to make one revolution) can be given a precise mathematical statement namely the square of the period is proportional to the cube of the distance, R, from the Sun to the planet $(T^2 = kR^3)$. (R is actually the length of the semi-major axis, which is approximately the same as the distance between the Sun and the planet because the ellipsity of the orbit is not very great.) This is Kepler's third law of planetary motion. His three laws using a more technical language may be summarized as follows:

**Law 1:** The orbit of every planet is an ellipse with the Sun at a focus.
**Law 2:** A line joining a planet and the Sun sweeps out equal areas during equal intervals of time.
**Law 3:** The square of the orbital period of a planet is directly proportional to the cube of the semi-major axis of its orbit.

## Origin of the Concept of Gravity

Kepler's three laws of planetary motion so crucial to Newton's formulation of classical mechanics describe only the motion of the planets and give no hint of the gravitational attraction, which keeps the planets bound to the Sun. Unless the Sun were exerting an attractive force on the planets they would fly off into space along a tangent to their orbit since according to the law of inertia a body will continue in constant speed straight-line motion unless a force is acting on it. The reason the planets do not fall into the Sun as a result of this attraction, which lies along the line joining the Sun and the planet, is due to the fact that a balance is struck between the Sun's attraction and the centrifugal force outwards due to the planet's motion. The net effect of the force is essentially to change the direction of the planet's motion without changing its speed so that a nearly circular orbit results. The tremendous contribution of Newton (and Hooke as well who made the discovery independently) was to realize that the force by which the Sun attracts the planets is the same as the one, which causes objects to fall to Earth. This presumably was the idea that dawned on Newton when according to legend the apple fell on his head.

The story of the modern development of the concept of gravity is an interesting one. It begins with Copernicus who believed that each of the heavenly bodies had its own local gravitational system. So in analogy to Empedocles idea that Earth attracted earth, water water, air air and fire fire, Copernicus believed that Moon attracts pieces of moon, the Sun pieces of sun and the Earth pieces of earth. In Copernicus' thinking there was no mutual attraction of the heavenly bodies and presumably a piece of moon would not fall to Earth but rather would fly up to the Moon and vice versa with a piece of earth on the Moon.

William Gilbert who published a book on magnets in 1600 played an important role in the development of the concept of gravity. He believed that the bulk of the interior of the Earth was lodestone and that magnetic attraction was the true cause of gravity. Gilbert believed the other planets, the Moon and the Sun also had magnetic properties and that these bodies mutually attracted each other like magnets.

Kepler was greatly influenced by Gilbert and incorporated these ideas into his cosmological system. He called the force emanating from the Sun effluvium magnetieum. Kepler's system did not include inertia. It explained the motion of the planets in terms of a positive force from the Sun, which presumably accounts for the fact that the closer a planet is to the Sun the faster it moves.

In the world systems of Copernicus and Kepler the space between the heavenly bodies is empty. In the Cartesian world system like the Aristotelian one, the space between the Sun and the planets is filled with a solid aether. Loberval, a follower of Descartes and the first to propose a universal gravitation attraction claimed in 1643 that the planets did not fall into the Sun because of the solid aether between them.

Borelli in 1665 came closer to the Newtonian picture by claiming that the planets did not fall into the Sun because the centrifugal tendency due to their circular motion was counterbalanced by the positive attraction of the Sun. With the suggestion of Borelli all the ingredients for the Newtonian breakthrough had been assembled. They included the Copernican heliocentric theory of the solar system, Kepler's three laws of planetary motion, Descarte's formulation of inertia, Borelli's concept of the centrifugal force as well as the concept of a universal gravitation attraction of all heavenly bodies. Descartes had also formulated the concept of momentum conservation, which he arrived at by arguing that God would not waste motion.

## Newton's Revolutionary Mechanics

Newton put the pieces of the puzzle together. Newton made two enormously important contributions. First he tied together all of the concepts that his predecessors had struggled to achieve and placed them within a consistent and coherent framework. Secondly he expressed the entire scheme in a precise, elegant mathematical language, which enabled one to calculate exactly the behaviour of mechanical systems. By assuming that the mutual gravitational attraction of two bodies is proportion to their masses and inversely proportional to the square of the distance between them, he was able to derive Kepler's three laws of planetary motion. He had realized the Pythagorean dream of expressing in terms of numbers the motion of the bodies of the universe. In so doing he completely transformed the nature of physics and the expectations of physicists who would never again be satisfied with anything less than a mathematical description of nature. Newtonian mechanics also influenced the non-scientist worldview ever after. But before explaining this let us consider the rudiments of Newtonian mechanics.

Perhaps the most important aspect of the Newtonian worldview is the way in which motion is regarded. For the early Greeks motion per se was not a natural thing and all motion had to be explained in terms of some force. Motion was often taken as evidence for the animation of the moving object. For the Newtonian on the other hand uniform straight-line motion (or constant velocity motion) is the natural state for any body and does not require an explanation. Any deviation from constant velocity, however, such as slowing down, speeding up or changing of direction must be explained in terms of some force. According to the Newtonian view once an object has attained a constant velocity in will retain that constant velocity until some force acts to change that velocity. We have all experienced this property of our own body, which we usually refer to an inertia. Perhaps you can recall being a standing passenger on an autobus and flying forward when the driver suddenly applied the brakes. By applying the brakes the driver produced a force, which slowed down the autobus. This force did not operate on you, however, and consequently you continued moving at the same speed, which is now greater than that of the bus and hence for an instant you traveled in the forward direction faster than the bus and hence experienced being thrown forward with respect to the bus. Actually what happened in that instant was that your motion remained the same and the bus was thrown backwards with respect to you by the force of the brakes. A similar thing happens when a passenger car suddenly turns to the left and we experience being pushed to the right. Once again our constant speed straight-line motion is unaffected by the force which makes the car change its direction. The car goes to the left and we go straight but with respect to the car we go to the right.

The principle of inertia serves as the foundation of Newtonian mechanics. Each particle is described by its mass, its position and its velocity. The mass of a body is a measure of the amount of matter it contains and is related to its weight. The actual weight of an object on Earth for example is a measure of the gravitational pull the Earth exerts on the object and is directly proportional to its mass. The same object on the Moon will weight less because the Moon's gravitational field is weaker than the Earth's. The mass of the object on the Earth and on the Moon is the same, however, even though their weights are different. The velocity of a particle describes both its speed i.e. the distance covered per unit time and its direction of motion. A body with a constant velocity is therefore one moving with a uniform rate of speed along a straight line. A planet orbiting a star in a perfect circular motion at constant speed does not have a constant velocity because it is continually changing its direction. Anybody whose velocity is undergoing change i.e. not remaining constant is said to be undergoing acceleration. The term acceleration in non-technical parlance usually means to speed up. For the physicist however the word has a more general meaning so that a body that is either speeding up, slowing down or changing direction is undergoing acceleration.

Although the maintenance of constant velocity motion requires no explanation in the Newtonian system all accelerations must be explained in terms of some force. The motion of a body does not change spontaneously; some other second body must be exerting a force to produce the change in motion of the first body. For example Newton explained the change in motion of the planets as they orbited the Sun as due to the gravitational attraction between the Sun and the planets or the acceleration of the proverbial apple that hit him upon the head as due to the gravitational attraction between the Earth and the apple. In these two examples the gravitational force is responsible for the change of motion. The force that one body exerts on another in order to change its motion, however, can also arise as a result of their colliding and transferring their momentum as occurs when two billiard balls collide and change direction.

Newton discovered in two body interactions, independent of the nature of the force, that the force that the first body exerts on the second body is equal and opposite to the force the second body exerts on the first. This concept is illustrated by considering a person diving from a raft. By pushing against the raft he dives forward into the water. The raft does not stand still but floats backwards. The force the raft exerts on the person propels him forward. The equal and opposite force of the person on the raft makes the raft float backward.

The same effect is illustrated by the fact that when a gun is discharged it recoils in the direction opposite to which the bullet is fired. The force of the bullet on the gun is equal and opposite to the force of the gun on the bullet. The reason that the speed of the bullet is so much greater than that of the recoiling gun is due to the large difference in their masses. The momentum of the gun's recoil and the bullet are equal and opposite, however. Momentum is the product of the mass times the velocity of a body. Momentum therefore has both direction and magnitude just like the velocity. In the example of the bullet and the gun the magnitude of the momentums are equal, their directions are opposite however and therefore when they are added they cancel. Both the example of the diver and the raft and the discharge gun illustrate the principle of the conservation of motion.

The magnitude of the momentum is a measure of how much inertia or motion a body is carrying. A very light object like a bullet is still very dangerous because if it is moving fast enough it carries a great deal of momentum. Of course if it travels at only one kilometer per hour there is no problem. A slowly moving object on the other hand, which is very heavy can also be dangerous. One does not wish to be caught between two freight cars even if they are only traveling at one kilometer per hour. The momentum is a measure of the amount of force a body can exert on others through collisions or as one of my students so aptly expressed it: "momentum is crushing power."

The transfer of momentum from one body to another is equivalent to the action of a force. In fact a force may be defined as an action of one body on another, which changes its momentum. Because the forces between two bodies are equal and opposite the total amount of momentum in any given interaction of two bodies is conserved. This is most easily observed by observing the elastic collisions of metal balls in the adjoining Fig. 6.3.

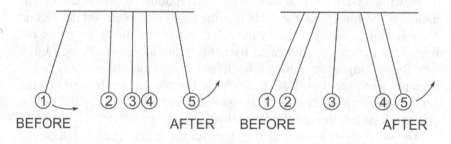

Fig. 6.3

If one ball falls on the others at rest momentum is transferred from ball 1 to 2 to 3 to 4 to 5, with the result that ball number 5 goes flying off with the same momentum as ball number 1. In second case the momentum is double because we let two balls fall, which transfers enough momentum, to allow both balls number four and five to go flying off with the same final momentum as balls number one and two had initially.

The conservation of momentum is a consequence of the fact that the forces between two bodies are equal and opposite. The two concepts are equivalent, as we have illustrated, by considering bodies, which have interacted with each other through collision or contact such as the bullet and the gun, the person and the raft, or the metal balls. Let us now apply these two concepts to the gravitational interaction where their validity seems less obvious. Superficial consideration of a rock falling off a mountain towards the Earth seems to contradict both the principle of momentum conservation and the idea that the forces between two bodies are equal and opposite. Before the rock falls there is no momentum but as a result of gravity the rock falls, develops velocity and hence momentum. It also appears that the Earth exerts a force on the rock but what about the equal and opposite force of the rock upon the Earth. The resolution of the paradox occurs when we recognize that actually the Earth is attracted to the rock and moves up to meet the rock in the same way the rock is attracted to the Earth and falls to meet it. Of course we never observe the Earth's motion, because the distance the Earth would move to meet the rock or the speed it would obtain as a result of this motion would be so small that it could not be detected but it nevertheless is there. Momentum is indeed conserved but since the mass of the Earth is approximately $10^{25}$ times greater than the mass of the rock, in order to conserve momentum its velocity is $10^{25}$ times smaller than the rock and hence unobservable. Nevertheless each time an object falls to Earth the Earth is falling up to greet the object in order to conserve momentum.

Another way of analyzing the Earth-falling rock system is to consider the relation between force, mass and acceleration According to the principle of inertia a body will move at constant velocity unless a force acts upon it, which is to say a body's acceleration is caused by a force. The more mass a body possesses the more difficult it is to alter its motion or produce an acceleration. It therefore follows that the force is equal to the product of the body's mass times its acceleration. This means that if two forces of the same strength operate on two bodies with different masses the lighter body will sustain a greater acceleration than the heavier body or putting it in terms of everyday experience it is easier to push a small rock than a big rock. In our previous example the force the falling rock exerts on the Earth and the force the Earth exerts on the falling rock are the same but since the mass of the Earth is so much greater than that of the rock the acceleration it experiences in unobservable. However, we can observe the effects on the Earth of another body, which is gravitationally attracted to it. This is the Moon, which is captured in an orbit about the Earth by the Earth's gravitational pull and which in turn pulls back on the Earth creating the high and low tides, which occur twice a day as the Earth rotates about its axis. The tides are a result of the Moon's gravitational pull on the waters of the oceans.

We have completed our discussion of the relationship of force and motion within the Newtonian framework. Newton formulated these ideas in terms of his famous three laws of motion, which we shall reproduce in their original form for historic interest. They also serve as a summary of our discussion.

**Law I:** Every body continues in its state of rest, or of uniform motion in a right [straight] line, unless it is compelled to change that state by a force impressed upon it.

**Law II:** The change in motion [rate of change of momentum] is proportional to the motive force impressed; and is made in the direction of the right [straight] line, which that force is impressed.

**Law III:** To every action there is always opposed and equal reaction; or the mutual actions of two bodies are always equal and directed to contrary parts.

The first law we recognize as the oft-cited principle of inertia. The first part of the second law literally states that the force equals the rate of change of the momentum. The momentum is the product of the mass times the velocity and, since the mass does not change, the change of the momentum is the mass times change of the velocity or the mass times the acceleration. Hence the first part of the second law is a statement that the force is the product of the mass times the acceleration which is often formulated mathematically as $F = ma$. The second part of Law II states that the change in motion occurs only in the direction of the force and embodies the principle of superposition. In other words if a car traveling at 30 m/s is driven off the side of a cliff it will continue to travel at 30 m/s, i.e. in the horizontal direction. Its velocity in the vertical direction, which was originally zero, will begin to increase once it is driven off the top of the cliff because of the pull of the Earth's gravitational field. Its motion in the vertical direction is unaffected by its horizontal velocity and it falls to the ground at the same rate as a car just pushed over the edge. The two cars would take the same time to hit the ground. However, the car originally traveling 30 m/s will have fallen the distance from the foot of the cliff that a car traveling at 30 m/s would have traveled on the ground from the foot of the cliff in the time it takes the car driving off the cliff to fall to the ground. Fig. 6.4 shows three cars, the one driven off the cliff at 30 m/s, the one, which just falls over the edge and an imaginary car traveling at 30 m/s on an imaginary road extended over the edge of the cliff. The time when each of the three cars reaches their respective positions is represented by their distance from the y-axis. Fig. 6.4 illustrates how the path of the car driven of the cliff at 30 m/s is a superposition of the paths of the two cars.

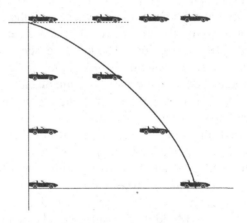

Fig. 6.4

Galileo formulated the principle of superposition in the following manner: If a body is subjected to two separate influences, each producing a characteristic type of motion, it responds to each without modifying its response to the other. This means if two or three forces act on a body the net effect of these forces is simply their sum. This is illustrated by considering the car sitting on top of the cliff. The force of gravity is pulling it down but the ground upon which it sits is also pushing up on the car with an equal and opposite force so the car does not fall, since the forces sum up to zero. Once it is pushed over the cliff the force of the ground pushing up can no longer act and so the car begins to fall.

The third law is a statement of the fact that whenever two bodies interact the forces they exert on each other are equal and opposite as was illustrated by the recoil of a discharged gun. Since the force acting on a body is a measure of the change of its momentum this law plus the principle of inertia insures that momentum will be conserved. The conservation of momentum may be stated as a law but it should be noted that it follows from Newton's three laws of motion and the definition of momentum.

Newton's three laws of motion offer a programme for describing the universe. If one knows the position and velocity of the particles in the universe at one given time and the mutual forces between them then one can predict the position and velocity of these particles for all future times. This is possible since knowledge of the forces implies knowledge of the accelerations. Then using the differential and integral calculus developed by Newton together with knowledge of the initial positions and velocities, one can use the information of the accelerations to predict all of the particles' future positions. The key to describing a mechanical system such as the solar system for example, therefore reduces to describing the forces between the various components of the system as well as their position and velocity at one given moment, which presumably one obtains through observation.

Indeed the first problem Newton applied his new mechanics to was the motion of the planets of the solar system. In order to solve this problem he had to make an assumption regarding the forces between the heavenly bodies. The notion of a universal gravitational interaction between the Sun and the planets, which also accounts for the Earth's local gravitational field had already been developed by Newton's predecessors, Copernicus, Gilbert, Roberval and Borelli who gave a qualitative description of the gravitational interaction.

In order to explain the interaction of the planets and the Sun using the mechanics he had developed, Newton required a quantitative understanding of the gravitational interaction. It was through his studies of Kepler's three laws of planetary motion that Newton came to understand the gravitational interaction quantitatively. He postulated that any two bodies would be mutually attracted to each other with a force directly proportional to the mass of each body and inversely proportional to the square of the distance between them. In other words the larger the mass, the stronger the force and the greater the distance, the weaker the force. When calculating the distance between two bodies to compute the strength of the gravitational force, one measures the distance from the center of one body to the center of the other as shown in Fig. 6.5. The mathematical description of Newton's law of gravitation attraction is $F = Gm_1m_2/R^2$.

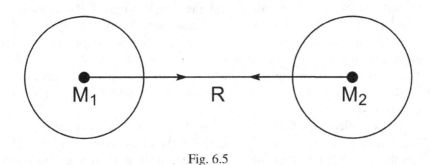

Fig. 6.5

Newton was able to show mathematically that only a force, which was inversely proportional to the square of the distance, could explain Kepler's law. However, the most convincing evidence he presented for his postulate was his comparison of the acceleration of an object at the Earth's surface with that of the Moon. Since the distance to the Moon was known, it was a simple matter to calculate the acceleration it experiences as it circles the Earth once a month. The acceleration of a body at the surface of the Earth is the same for all bodies as will be explained below and is easily measured. When one compares this acceleration with the lunar acceleration and the distances from the center of the Earth to its surface and to the Moon, one easily verifies that the gravitational force is inversely proportional to the distance between the two bodies squared.

Newton's result depended on the fact, which had been experimentally verified, that the acceleration a body undergoes as a result of the Earth's gravitational pull, is independent of its mass. In other words, the fact is that a five-pound stone falls as rapidly as a fifty pound stone. One can cite a counter example such as the comparison of a stone and a feather, in which case the stone falls faster than the feather. This example is more complicated than the example with the two stones because the air resistance encountered by the feather produces an effective force, which retards it descent to Earth, which is a negligible effect with stones. If the same experiment were repeated in a vacuum then indeed the feather and the stone would fall at exactly the same rate in the vacuum. The explanation for why all masses fall at the same rate when air resistance may be ignored is due to the fact that the force acting on the falling body is proportional to its mass and that this force produces an acceleration inversely proportional to the falling body's mass ($F = ma$). As a result the effects of the mass cancel and the acceleration of the particle is independent of its mass. Putting it less technically, the greater the particle's mass, the greater the gravitational force it experiences. The greater its mass, however, the more difficult it is to change its velocity. These two effects of the mass just cancel so that all bodies accelerate in the Earth's gravitational field at the same rate as is verified by experiment.

If we reflect carefully on the role the mass of a particle plays in Newtonian physics we quickly realize that the mass serves two functions. On the one hand it determines the strength of a body's gravitational interaction (the greater the mass of a body the stronger is its gravitational interaction). On the other hand the mass also determines the body's resistance to acceleration or a change of its velocity. The acceleration a body experiences as a result of a force is inversely proportional to its mass since the force is equal to the product of the mass times the acceleration ($F = ma$ or $a = F/m$). Hence the greater the mass of a body, the smaller the acceleration it experiences as a result of a given force. The mass also determines the momentum or amount of motion since the momentum is the product of the mass times the velocity. The greater the mass of a body the greater its momentum for a fixed velocity and hence the more difficult it is to change that velocity.

The property of a body to attract other bodies gravitationally and the property to resist acceleration are two separate properties of a body, each ascribed to the body's mass. However, these two properties are so different that we should recognize that a body has really two masses; a gravitational mass, which generates a gravitational force with other masses and an inertial mass, which resists being accelerated when a force is applied to it. These two masses, which are quite distinct, nevertheless turn out to be equal. This seemingly bizarre coincidence accounts for the fact that all masses fall at the same rate. The heavy bodies experience a greater gravitational pull. However, they are less affected by the force because of their greater inertial mass. Because of the equality of these two masses for all bodies, these two effects always cancel so that all masses fall at the same rate.

The equality of the gravitational and inertial masses could be a coincidence. However, it most likely reflects a deeper relationship between the two masses. Ernest Mach in the nineteenth century suggested that the inertial mass of a body arises as a result of the gravitational pull of all the other bodies in the universe. According to this hypothesis if a body found itself alone in the universe it would not have any inertial mass. The properties of a body depend on all the other bodies. All things are interconnected. This idea is incorporated in Einstein's General Theory of Relativity to be discussed later. But even Newton's formulation of gravity and mechanics contains this concept. Every particle of matter in the universe is gravitationally attracted to every other particle of matter. This means that you and I are at every moment in contact with every piece of matter in the universe. We are pulling every element of the universe towards us and in turn are being pulled toward each particle of the cosmos. Of course the matter composing the Earth has the greatest influence upon us because it is closer to us than the other masses in the universe.

Consideration of tides reveals how a body 400,000 kilometers away (the Moon), can still exert an influence upon us. It is almost like magic — the gravitational attraction of matter. Let us consider the Earth and the Sun separated by 150 million kilometers of empty space and yet exerting an action upon each other. The reason I describe this interaction as magic is that if you want to exert a force on a friend, push them in some direction for example, you would consider some way of making contact with them. Either you would touch them directly with your hands, or with some object in your hands such as a pole, or perhaps you would throw something at them to force them to move in some direction. If you could physically make your friend move by merely looking at them without making some type of physical contact, you would be considered a magician. And yet, we accept the idea that the Sun can affect the motion of the planets without touching them, i.e. produce an action at a distance. There is a magic there alight.

There is no way we can explain this action at a distance property of the gravitational interaction other than to say this is the way matter behaves. The only serious criticism made of Newton's theory of motion and gravitation was of this principle of action at a distance. Newton, however, never claimed to have explained gravity. He only described it. So long after we have described planetary motion in terms of inertia and gravity, we are still left with the question of how do two masses separated by a distance with nothing but empty space between them, exert a pull on each other. The only answer we can give is they do because they do. This is the way our universe is constructed. Can you think of a better way? I can't. Physics only describes the universe in terms of a small number of principles or laws but it cannot explain why these basic laws are the way they are. There are certain basic laws or principles upon which physicists build their description of the universe and all we can do is to describe these laws or principals. Explaining why they exist or why the universe exists are not questions amenable to scientific analysis.

## Chapter 7

# Poetry Influenced by the Scientific Revolution

In this chapter, we plan to trace the influence, the rise of modern science had upon the thinking in other areas of human enterprise and thought. By the rise of modern science we are referring to the revolution in human though brought about by Copernicus, Kepler, Galileo, Newton and their co-workers, whom we have just finished discussing in the previous chapter. This period of scientific work covers roughly the period 1500 to 1700.

Before the scientific revolutions of this period, thought in Europe was dominated by religious matters. During the Middle Ages the only order imaginable was divine order and all events were understood as a reflection of this order.

During the Renaissance there was a shift of emphasis in which thought became man-centered rather than strictly God-centered. Educated people became interested in scientific knowledge but they still place their ultimate salvation and trust in their knowledge of God. Even the scientists, who produced the scientific revolution, including Copernicus, Kepler, Galileo and Newton were all deeply religious men who believed that their discoveries revealed more deeply the glory of God. The Renaissance attitude toward science and religion is perhaps best reflected in the poetry of John Donne, who incorporated scientific ideas into his verse. In his poem The Ecstasy he makes use of the concept of atoms

> We then, who are this new soul, know,
> of what we are compos'd, and made
> For, the atomies of which we grow,
> are souls, whom no change can invade.

In his poem A Valediction: Forbidding Mourning, he uses the compass as a simile referring to the souls of his wife and himself;

> If they be two, they are two so
> As stiffe twin compasses are two
> Thy soul the fix't foot, makes no show
> To move, but doth, if the other does.
> And though it in the center sits
> Yet when the other far doth roam
> it leaves, and hearkens after it
> And grows erect, as that comes home.
> Such wilt thou be to me, who must
> Like its other fool, obliquely runne;
> Thy fermnes draws my circle past
> And makes me end, where I begin.

In his poem Love's Alchymie, he compares the hopes of two lovers' with those of an alchemist's. With regard to the issue of the heliocentric versus geocentric universe his poetry clearly indicates his acceptance of the heliocentric view albeit in the form of Tycho Brahe's compromise system. In spite of his acquaintance and acceptance of scientific ideas Donne's poetry still retains a skeptical attitude with regard to the value of scientific knowledge. For him the most important knowledge is that of God. He would argue that scientific ideas will come and go but a belief in God will endure. This attitude is reflected in the following passage of the Second Anniversary;

> Why grass is green, or why our blood is red
> Are mysteries which none have reach'd into.
> In this low form, pour soul, what will thou do?
> When wilt thou shake off this pedantery
> Of being taught by sense and fantasy?
> Thou look'st through spectacles, small things seem great
> Below, but up onto the watchtower get
> And see all things despoil'd of fallacies
> Thou shalt not peep through lattices of eyes,
> Nor hear through labyrinths of ears, nor learn
> By circuit or collection to discern.
> In heaven thou straight know'st all concerning it.
> And what concerns it not, shalt straight forget.

A change in attitude towards science began to develop in the beginning of the seventeenth century during the period in which Galileo, Kepler and a number of other scientists were uncovering the mysteries of nature, which eventually led to Newton's great breakthroughs. This new attitude was reflected in the philosophical writings of three great thinkers of this period, namely Francis Bacon (1561–1626), Thomas Hobbes (1588–1679), and Rene Descartes (1596–1650).

Francis Bacon was one of the first philosophers to extol the virtues of science. He was a strong advocate of the empirical approach. He felt that it was solely through the observation of nature that all knowledge would be discovered. His view was a trifle naive because he did not properly take into account the role deductive thinking would play in physics. His influence in stimulating experimental work, however, certainly made a positive contribution to the scientific activity of his day. He also accurately predicted the central role that science would come to play in the life of humankind and the importance that joint research projects would play.

Hobbes is the first thinker we encounter not directly involved in scientific activity like Bacon and Descartes, whose philosophy is nevertheless intricately connected with the new science. He was a great admirer of mathematical thought, being particularly interested in geometry. He was also a thorough going empiricist, on the other hand, with a healthy respect of the inductive method. Philosophically he addressed himself to political questions. A pessimist concerning human nature, his philosophical system presents a drab deterministic and materialistic view of life. His description of humans and their institutions was extremely mechanical, not unlike our stereotyped picture of how a mad scientist, like the movie character Dr. Strangelove, regards life.

Descartes, the third philosopher under consideration was also a mathematician and a physicist. His greatest contributions were in mathematics and philosophy. His philosophy was greatly influenced by his scientific thinking. He adopted a position of skepticism, never accepting any philosophical truth from the past that he could not verify for himself. This spirit of skepticism, known as Cartesian doubt, is identical to the spirit in which modern science is conducted. It become one of the cardinal principles of modern philosophy and is perhaps science's greatest contribution to the development of philosophic thinking or, perhaps vice versa, philosophy's greatest contribution to science.

We have so far traced the development of the influence of scientific thinking on the humanities prior to the appearance of Newton's Principia Mathematica in 1687. The work of Newton brilliantly completed the scientific revolution begun by Copernicus. He was an intellectual hero of his time as the following lines of Pope reveal:

> Nature and Nature's law lay hid in Night!
> God said, "Let Newton be!" and all was Light.

Newton's laws of motion explained both the movement of the planets about the Sun and that of projectiles in the Earth's gravitational field. As a result his work was almost immediately accepted by both the scientific and lay communities. His precise mathematical description of both heavenly and earthly motion had a tremendous effect on subsequent thinkers in all fields of human endeavor. The entire world now seemed capable of being described rationally, mathematically and mechanically. All were eager to imitate Newton's success in the description of all phenomena including such diverse things as human behaviour. Mechanical models were used to describe everything from the solar system to the human mind, from the economics of the market place to the creation of the universe by the deity.

This new attitude towards science is probably best expressed by the poet John Dryden (1684) in his essay Of Dramatic Poesie in which he wrote:

> Is it not evident in these last hundred years (when the study of philosophy has been the business of all the Virtuosi in Christendom) that almost a new Nature has been revealed to us? — that more errors of the school have been detected, more useful experiments in philosophy have been made, more noble secrets in optics, medicine, anatomy, astronomy discovered, than in all those credulous and doting ages from Aristotle to us? — so true is it that nothing spreads more rapidly than science, when rightly and generally cultivated.

We devote the remainder of this chapter to sketching the effect of the Newtonian revolution in physics on the thinkers in other fields of human endeavor. One of the great impacts of the new science was the influence on religious thinking. The scientists themselves such as Copernicus, Galileo, Kepler, Boyle, Hooke and Newton were rather orthodox in their religious thinking or at least they claimed to be. This is a fairly safe surmise in the case of Newton who considered his dating of events in the Bible as important as his work in physics. Newton believed that the order in the universe that his physics theories had uncovered revealed the greater glory of God and reaffirmed his faith in the Christian deity. According to orthodox Christian thinking, however, God often intervened in worldly matters and is actually actively engaged in the running of the universe from day to day. To many thinkers this seemed incompatible with the Newtonian world picture in which the objects of the universe behave predeterministically according to certain well-defined mathematical equations. To these thinkers it was natural to relegate to God solely the job of creation of the universe, which once created would run according to His law. There was no need for God's intervention. Voltaire used the analogy of the clockmaker and his clock to describe the relation of the Deity and His universe. He wrote, "I cannot imagine how the clockwork of the universe can exist without a clockmaker."

This religious viewpoint referred to as Deism, perhaps first encountered in the thinking of Bruno, still involved a worshipful attitude toward the deity as is evidenced by Addison's ode:

> The spacious firmament on high,
> With all the blue ethereal sky,
> And spangled heavens, a shining frame,
> Their great Original proclaim:
> The unwearied Sun, from day to day,
> Does his Creator's power display,
> And publishes to every land,
> The work of an almighty Hand.

Other thinkers took the Deist position to its logical conclusion of a universe without a deity governed by mechanical and mathematical natural law, a position similar to that of Thomas Hobbes.

Perhaps the greatest impact Newton physics made outside the field of science itself was on the field of philosophy. Two major philosophical movements, one in England (the British Empiricists consisting of Locke, Berekeley and Hume) and the other in France (the Philosophes consisting of Voltaire, Condillac, Diderot, Condoret and others) based their philosophical methods directly upon the scientific methods, which formed the basis of the Newtonian system. These thinkers attempted to answer philosophical questions including those concerning the working of the human mind through the observation and analysis of empirical information in direct analogy to the methods of physics. This attitude is perhaps best expressed by the following remark of Hume: "As the science of man is the only solid foundation for the other sciences, so the only solid foundation we can give to this science itself must be laid on experience and observations."

The British Empiricists by adopting the methods of science converted the study of humankind from philosophy to psychology, a field of study that perhaps first commences with their work. Although they were a trifle naive by present day standards, employing rather crude mechanical models for the way in which the human mind worked, they established the empirical foundations of the study of psychology.

The British Empiricists like the Pre-Socrates believed that knowledge is arrived at only through the observations made by the body's senses of sight, hearing, feeling, smelling and tasting. They attached different values to the various senses, however, and categorized the properties of an object apprehended by the senses, into primary quantities of an object consisted of extension, position, motion, mass and density, whereas the secondary quantities consisted of color, sound, taste and odor.

The primary quantities are just the properties of a body that are necessary to describe it within the framework of Newtonian Physics. The secondary quantities, on the other hand, are not needed for a mechanistic description of a body. Furthermore, the primary quantities are explained i.e., related to each other within the framework of Newtonian mechanics whereas at the time these philosophers were writing, the secondary quantities such as color, sound, odor and taste were not understood scientifically as were the primary quantities.

As a consequence, primary and secondary quantities were regarded differently by these philosophers. The primary quantities were held to be susceptible to rational analysis and to have an objective reality whereas the secondary quantities were held to be purely subjective, not even a property of the body but rather of the observer. In other words what can be understood is real and what cannot be understood is not real and hence unimportant.

This attitude of the empiricist is similar to that of certain Greek thinkers like Plato and Aristotle, who divided the world into two domains; one domain, which they understood and called real and rational and another domain they did not understand and labeled unreal and irrational. For Plato, it was the world apprehended by the senses that was unreal and irrational, whereas the world of ideas was the ultimate reality, where truth and understanding were possible. Aristotle, on the other hand, divided the universe into the heavens and the Earth. The heavens were unchanging and rational whereas the Earth was constantly changing, unexplainable and irrational. This dichotomy also found its way into Christian theology. It has in fact remained a feature of Western thought, which persists in contemporary thinking. The only change in this mode of thinking has been a shift of more and more material from the unreal irrational compartment to the real rational one as our understanding of nature increases.

Today in our technocratic society, it is the questions that we are able to address with our science and technology that is the ultimate reality and hence is given first priority. The more complicated problems involving human emotions, the existence of which many technocrats try to deny, are often ignored or treated in an extremely superficial manner.

During our discussion of the roots of scientific thinking we suggested that the legal codes employed by the Greeks in organizing their political life, influenced their concept of natural law. With Newton's discovery of his laws of motion, a reversal of this connection took place in which the concept of natural law influenced philosopher's concept of political law. People began to think of government differently. They were looking for a legal system in harmony with the natural laws of human behaviour, a position that was epitomized by the works of Jean Jacques Rousseau. Others such as Locke and Hume in England, the Enlightenment philosophers in France and Jefferson and Paine in the United States began to question existing political institutions. They applied scientific principles in their quest for new answers and new models of political organization. The political consequences of this intellectual ferment were far-reaching and culminated in the American and French revolutions.

The American constitution with its systems of checks and balances is a reflection of the mechanistic models political thinkers were employing as a result of Newton's influence. The scientific method was also applied to economic questions and lead in the example of Adam Smith to his formulation of his economic theory in terms of the law of supply and demand.

The blessings of the Newtonian revolution in physics were mixed. There is no doubt that there was a sudden and dramatic increase in man's understanding of his universe. The progress in the scientific fields was quite obvious. The application of scientific methodology to the humanities particularly philosophy and politics, was, on the whole quite positive. There were excesses, however, in which scientific methods were applied to humanistic problems in an uncritical manner. This is perhaps best illustrated by the extreme materialistic doctrine of certain members of the French philosophy movement such as La Mettrie and his disciple Cabanis. La Mettrie was a complete and total materialist who believed that every human property, characteristic and activity could be completely accounted for in a purely physical and mechanical manner. He wrote:

> Man is a machine so compounded that it is at first impossible to form a clear idea of it, and consequently to define it. That is why all the investigations, which the great philosophers have conducted a priori, that is to say by trying to lift themselves somehow on the wings of their intellect, have proved vain. Thus, it is only a posterior or by seeking to unravel the soul, as it were, via the organs of the body, that one can, I do not say lay bare human nature itself in a demonstrative fashion, but attain to the highest degree of probability possible on this topic.

Cabanis, who was a physician like La Mettrie, proposed that the brain secretes thought as the liver secretes bile. These very first behaviourists denied the existence of the human spirit in the same way the advocates of strong artificial intelligence do today. It was, in part, against this type of uncritical abuse of scientific thinking that the reaction known as the Romantic Movement took place.

The Romantic Movement was a reaction against the Enlightenment in general. The romantics believed that reason divorced from feelings and emotion, which in general characterizes scientific thinking, led to disastrous results such as the Industrial Revolution, which had given rise to grave social problems. Still another cause for criticism of science was the social disorder that grew out of the French Revolution; a revolution spawned by the rationalistic philosophy that in turn was inspired by the scientific revolution, which preceded it by one hundred years.

Although the Romantic Movement occurred wherever science was studied, it was in Germany where the movement really flourished. Goethe perhaps the leading proponent of romanticism, summarized the movement's attitude towards science with the lines

> gray and ashen, my friend is every science,
> and only the golden tree of life is green.

Although the romantics were disillusioned with science they were not disillusioned with the subject of scientific study, namely nature, as the following lines of Blake indicate.

> To see a World in a grain of sand
> And a Heaven in a Wild Flower
> Hold Infinity in the palm of your hand
> And Eternity in an Hour.

The spirit contained in these lines does not represent a critique of science but rather an implicit support of it. The sentiment expressed could well be that of a dedicated scientist. It has always seemed to me that the conflict between the scientists and the humanists, epitomized by the critical attitude of the Romantic Movement toward science, is without substance.

The criticism of the humanists is often quite correct, but they are often misdirected because they do not distinguish between science and pseudo-science. The science community on the other hand, has been lax in criticizing those who abuse either their methods (the pseudo-scientists), or their results (technocratic exploiters of humanity). Hopefully the day will come when these two communities can combine their talents so that man's knowledge will no longer be used destructively against his own self interest as it has been at various times since the advent of scientific revolution such as at Hiroshima, or in the sweat shops spawned in the early days of the Industrial Revolution.

# Chapter 8

# The Concept of the Atom, the Atomic Structure of Matter and the Origin of Chemistry

The internal structure of matter cannot be perceived directly with our senses. The surface of this page seems to be smooth and whole. The material of which it is composed appears to be static and unchanging. All of this is a deception.

An examination under an extremely powerful microscope would immediately reveal an incredible amount of structure on the surface of the page. In fact, if we were able to reduce ourselves in size such that the area of this page became equivalent to the surface area of the Earth, we would discover that this smooth piece of paper possessed mountains ten times the height of the Alps and canyons ten times deeper than the Grand Canyon.

Let us for the moment consider a ball rolling on the surface of this page. Its surface is likewise covered with gigantic mountain ranges and canyons. From this vantage point it is easy to see how the force of friction operates. As the mountains of the rolling ball and the paper rub against each other the motion of the ball is interfered with and it slows down just as a ball rolling on a cobble stone road slows down faster than one rolling along a paved road.

Let us return to our examination of this seeming solid and static sheet of paper upon which these words are printed. If we were to reduce our size once again so that the area of the page now becomes equivalent to the area of the Earth's orbit about the Sun, we would become aware of the atomic structure of matter. What a surprise to discover that this static sheet of paper is a maze of motion. Not only are all the atoms furiously oscillating back and forth with respect to each other, but also each one is a center of internal motion. Within each atom electrons are whizzing about their respective nuclei, imitating the planets of a miniature solar system. The force holding the electrons in their orbit is not gravity but the electric force of attraction between the proton's positive charge and the electron's negative charge. Even the protons and neutrons tightly bound within the nucleus are moving back and forth with respect to each other. Nothing is standing still in this "Restless Universe" (the title of an excellent book on the atom by one of the pioneers of quantum physics, Max Born).

Not only is our solid page a maze of motion, but also it consists almost entirely of empty space. The distance between atoms and the radius of the electron's orbit about the nucleus are approximately the same namely $10^{-8}$ cm. The radius of the nucleus on the other hand is only 1/100,000 of this distance or $10^{-13}$ cm. The radius of the electron is even smaller. Actually, according to some theories an electron is a geometric point. Thus like our solar system the atom consists mostly of empty space and hence so-called solid matter is 99.999 ...% empty space. In fact if it were possible to compress matter such that all the nuclei were touching each other so there was no longer any empty space within the atom, then a library consisting of $10^{15}$ books the size of this one could be compressed into the space occupied by this book. However, this is pure fantasy since the force here on Earth, necessary to compress matter to this extent is impossible to achieve. The existence of such super dense collections of matter cannot be ruled out, however, as we will discover later in the book, when we consider cosmological questions and investigate black holes.

Having briefly sketched the nature of the atomic structure of matter we will leave the detailed discussion of this topic to the later chapters when we describe the physics that takes place inside the atom. Instead we now turn our attention to the implication of the atomic structure of matter on its macroscopic (large scale) behaviour. The concept of an atom invisible to the human eye and immutable was first proposed by Leucippus and Democritus to account for the impermanence of matter. According to them, although the nature of atoms is unchanging, changes occur on the macroscopic level of matter as different combinations of atoms form and then re-form. So for example, as wood burns the atoms of wood and air form new combinations such that fire, ashes and smoke result. They believed that the space between atoms was void and that the properties of matter were determined by the size and shape of the atoms as well as the nature of their motion in the void between them. Their ideas, never very popular, were rejected by Plato and Aristotle outright, because they involved the concept of a void. The Greek philosopher Epicurus adopted the atomists' point of view as the basis of his system of thought. The Latin poet Lucretius preserved the teachings of Epicurus through his epic poem, De Rerum Natura (On The Nature of Things). It was through this work that the thinkers of the Renaissance became acquainted with the ideas of the atomists.

> This ultimate stock we have devised to name
> Procreant atoms, matter, seeds of things,
> Or primal bodies, as primal to the world.
> (Proem — Book I of On The Nature of Things)

Perhaps the most influential Renaissance supporter of atomism was the English philosopher and scientist Francis Bacon (1561–1626), the great proponent of the empirical approach. He was the first to suggest that the heat of a body is a result of the internal motion of its atoms. The atomism of Bacon was identical to that of the Greeks in all its details except for one point, which reflected the growing rationalism of Bacon's time. The Greeks believed that the combining and recombining of the atoms was a random process governed by chance whereas Bacon and his followers believed that this process was governed by rationalistic law. The mechanical nature of these atomic processes was emphasized by the English physicist and chemist Robert Boyle (1627–1691). He discovered that for a given quantity of gas at a fixed temperature the pressure is inversely proportional to the volume. Although this result can be derived theoretically from an atomic picture of gases, Boyle discovered his law through experimentation. Other physicists such as Hooke, Galileo, and Newton also believed in the existence of atoms.

In opposition to the view of the modern atomists were the Cartesians who adopted the view that matter was continuous and could be divided and subdivided an infinite number of times without ever reaching an end to the process. Part of the Cartesians' objection to atomism was their opposition to the concept of a vacuum. They believed that even if one could remove all the air from a tube using a pump, that aether would remain to fill the empty space. They believed also that the space between the heavenly bodies was also filled with aether, a point of view that survived until Einstein's relativistic interpretation of the Michelson-Morley experiment, which we will discuss in Chapter 13.

In the absence of any empirical information regarding the structure of matter, the Baconian-Cartesian controversy over the existence of atoms was for a long time a philosophical one. The resolution of this question of physics was finally achieved by a chemist. John Dalton, whose interpretation of certain empirical laws of chemical combination provided the first observational evidence for the existence of atoms. Dalton's work is the first example we shall encounter, which demonstrates the intimate connection between chemistry and physics. This should be no surprise for according to our definition of physics as the study of nature, it is obvious that the distinction between physics and chemistry is artificial and is a result of historic accident. The same may be said of the division between astronomy and physics. Just as Newton's theory of motion and gravitation displayed the unity of astronomy and physics so has the concept of the atom united chemistry and physics.

Chemistry grew out of the art of alchemy, a practice that has been somewhat maligned in modern times. The alchemists attacked the problem of converting the base metals into gold. The reason their practice has been maligned is that a number of the assumptions that formed the basis of their work have subsequently been shown to be incorrect and hence the alchemists were engaged in a futile attempt to convert the base metals into gold. They believed like many of the early Greeks and Mesopotamians that earth, water, air and fire were the basic elements out of which all things in this world are made. To this list they added two other elements — mercury, which they claimed imparts metallic qualities to substances and sulfur, which produces inflammability. They regarded metals as compounds of the elements fire, and earth, a natural assumption on the basis of their observations. Metals such as iron and copper were produced by reducing metallic ore in a very hot fire, hence they thought a metal was a compound of mercury and earth (the ore) and fire. In actuality iron ore is a compound of iron, oxygen and impurities, which under the conditions of extreme heat reduces to pure iron. This was unknown to the alchemist, however, who hoped that by adding more fire or perhaps mercury and sulphur or all three to a metal like lead or iron they would be able to produce gold.

It is unfair to be critical of their attempt to alter the metallic elements chemically when they did not realize that the metals of iron, lead, mercury and copper were basic elements. Although they never succeeded in obtaining their goal they established the experimental foundation of modern chemistry. In fact the boundary between alchemy and chemistry is an artificial one created by textbook writers. They ascribe those results, which still stand today as due to the work of chemists and those results, which have been discarded as due to alchemists. This is absurd if we recall that Robert Boyle the so-called father of chemistry believed that gold was a compound that could be manufactured from other metals as he described in an essay entitled *Of a Degradation of Gold made by an anti-elixir: a strange chymical narrative,* which was published in London in 1678. The great Isaac Newton, the inaugurator of modern mathematical physics, was also an alchemist. Because of certain fraudulent practitioners of alchemy, alchemy got a bad reputation so that the genuine scientific practitioners rebranded their science and called it chemistry.

Although the alchemists believed that all chemical substances were combinations of certain basic elements it was not until 1661 that Robert Boyle established the modern definition of an element and a compound. A compound is a substance that can be broken down into simpler elements whereas an element is a substance that cannot be broken down any further. Although Boyle defined the concept of an element and a compound, subsequent research revealed that a number of substances he identified as compounds were elements and vice versa. By 1800 as a result of the work of such noted chemists as Lavoisier, Davy, Scheele, Black, Priestly and Cavendish, over 30 elements had been correctly identified, and a large number of chemical reactions had been studied quantitatively and the qualitative distinction between a process of mixing and a chemical reaction was discovered.

A mixture retains the qualities of the substance of which it is composed. A chemical compound, on the other hand, is completely different than its components. For example, the two gases (hydrogen and oxygen) combine to form a liquid (water). The poisonous green gas chlorine combines with the highly reactive metal sodium to form the harmless substance, sodium chloride, we recognize as table salt.

It was found that mixtures could be prepared with any ratio of its components. With a chemical compound, on the other hand, the ratio of the weight of its constituents is always the same. So for example, it was found that one gram of hydrogen combines with exactly eight grams of oxygen to form 9 grams of water ($H_2O$). It was also found that if two elements, A and B, formed more than one compound, then the amounts of the element A, which combined with a fixed amount of B are related to each other as the ratio of whole numbers, which are generally quite small. Hydrogen and oxygen also combine to form hydrogen peroxide ($H_2O_2$), with one gram of hydrogen combining with 16 grams of oxygen. The ratio of the amount of oxygen combining with a fixed amount of hydrogen for hydrogen peroxide and water is therefore two to one.

From these two laws of chemical combination referred to as the law of definite proportions and the law of multiple proportions, John Dalton in 1808, deduced that all matter is composed of elementary particles, which he called atoms after Leucippus and Democritus. He concluded like the early Greeks that these atoms retain their identity in chemical reactions. He also deduced that all the atoms composing an element are identical and that compounds consist of identical molecules. The molecules are themselves aggregates of atoms, which have combined in simple numerical proportions.

So, for example, a water molecule consists of two hydrogen atoms and one oxygen atom, $H_2O$ (H and O are the symbols for hydrogen and oxygen), whereas hydrogen peroxide molecules consist of two hydrogen atoms and two oxygen atoms, $H_2O_2$, which explains why twice as much oxygen combines with a fixed amount of hydrogen to form hydrogen peroxide as opposed to water. Once it was known that for water there are two hydrogen atoms for each oxygen atom the relative weight of the oxygen atom and the hydrogen atom was determined to be 16 to 1. By considering all of the various reactions the relative weight of all the atoms were obtained. In the very first compilation made by Jons Berzelius in 1828, a scale was chosen such that the atomic weight of hydrogen was one. This scale has been refined so that today carbon has the atomic weight of exactly 12.0.

From the fact that two volumes of hydrogen gas plus one volume of oxygen gas combine to form two volumes of water vapor instead of one volume, we learn that both hydrogen and oxygen gas consists of molecules containing two hydrogen and two oxygen atoms respectively. The other gaseous elements with the exception of the noble gases like helium and neon are also diatomic.

# Chapter 9

# The Concept of Energy

In the last chapter, we showed how the concept of the atom, which has proven so valuable to physicists, was developed basically by chemists. The concept of energy to be dealt with here is another concept playing a central role in physics, which was developed partially through the efforts of the chemists, further illustrating the point that the division between physics and chemistry is arbitrary.

The concept of energy has always been associated with the idea of its conservation. The origins of this idea most likely originated with the conservation of mass implicit in Newton's equations of motion. This assumption was strictly limited to mechanical reaction for it was thought that for certain chemical reactions, such as burning, mass was not conserved. This misconception was due in part to a lack of an understanding of the process of oxidation, which was thought of as a process whereby the burning object released a substance called phlogiston, a word derived from ancient Greek, which meant "burning up" and in turn was derived from the ancient Greek word phlox, which meant fire. The theory first postulated by Johann Joachim Becher in 1667 postulated that phlogiston had a negative weight to account for the fact that the products of combustion were heavier than the original substance, which burned. Lavoisier on the other hand, correctly believed that burning was due to the oxidation of the burning substance and that the increase in weight was due to the weight of the oxygen that combined with the burning substance. He proved this by carrying out oxidation in a completely closed system and showed that the total amount of mass before and after combustion was the same. He therefore postulated the conservation of matter held for all reactions including both chemical and mechanical ones.

Lavoisier's conservation of mass did not, however, lead directly to the conservation of energy but provided a model for it. It also provided a model for the erroneous concept that heat is a conserved quantity. Heat was considered to be weightless fluid called caloric. The transfer of heat from a warm body to a cool body involved the flow of the fluid caloric, which was conserved. The generation of heat as a result of friction would appear to contradict the idea that heat was a conserved fluid.

The generation of this heat was explained however, as arising from caloric being squeezed out of the body by the action of the friction. This would mean that only a fixed amount of heat could be generated from any given body due to friction for once all of the caloric had been squeezed out of the body, it would have no more heat to give. Count Rumford (nee Benjamin Thompson) in 1790 or thereabouts, while working in a cannon factory observed a direct contradiction of this idea. He noticed that an inordinately large, almost unending, amount of heat was generated in boring the hole necessary to convert a large metal rod into a cannon.

Further study quickly revealed that he could generate as much heat through friction as he wanted, as long as he provided the necessary work to generate the friction. You can perform this experiment yourselves by rubbing your hands together. The only limitation on the amount of heat you create will be the physical exhaustion you experience from rubbing your hands together. Count Rumford discovered that the amount of heat is not conserved and that the creation of heat requires work.

Heat is also generated when a moving object suddenly comes to rest. For example, if a ball falls to Earth, the temperature of the ground and the ball increase immediately after the ball strikes the ground and comes to rest. In both the example of friction and the falling ball motion is converted into heat. An understanding of these processes involves the realization that the heat of an object is nothing more than the internal motion of the atoms of which it is composed. Motion is converted into heat simply because the external motion of the object is converted into the internal motion of its atoms. Work is converted into heat by first creating motion, which in turn is changed into heat. The connection between work and motion is quite obvious. It takes work to create motion. A horse must work to pull a cart. It takes work to lift an object to a certain height in order to give it motion by dropping it.

The above discussion illustrates the equivalence of heat, motion and work. It is the concept of energy that ties together these three quantities, for it requires energy to perform work, to create motion or to generate heat. The term energy comes from ancient Greek word energos, which meant "active working".

The conservation of energy is nothing more than the statement that heat, motion and work are equivalent and that for a given amount of work one gets the same amount of motion or the same amount of heat, or that for a given amount of motion one gets the same amount of heat and so on and so forth.

It requires energy in the form of work to give motion to a body, which is initially at rest. The energy acquired by a body in motion is referred to as its kinetic energy. Since all objects are composites of smaller particles called atoms, which are also in motion, the motion of any object can be separated into its external motion and its internal motion. The term, the kinetic energy of a body, usually refers to the energy due to its external motion. The energy of its internal motion, on the other hand, is by definition heat. The amount of heat is exactly equal to the sum of the kinetic energy of each atom's internal motion. The amount of energy required to move an object depends on its mass and the final velocity of its motion. The greater either the mass or the velocity, the greater the energy. Since energy was defined as a conserved quantity, the kinetic energy of a body was defined equal to one half its mass times its velocity squared ($E = 1/2\ mv^2$) to insure that energy remains a conserved quantity. Kinetic energy is not necessarily conserved as was illustrated by bodies that slow down as a result of friction or the ball that came to rest as a result of striking the Earth. In each of these examples the kinetic energy is converted into other forms of energy such as heat in the case of friction or in the case of a ball striking the Earth kinetic energy is converted into heat, sound and a deformation of the ground.

As mentioned in our previous discussion of mechanics, a force is necessary in order to change the velocity a body. In order to exert a force however, energy must be provided in the form of work. The amount of work done as a result of exerting a force is equal to the distance through which the force acts, times the magnitude of the force in the direction through which it acts. If a force acts perpendicular to the motion of a body no work is necessary since this action will only change the direction of the body and hence the kinetic energy will remain fixed. If the body is pushed along its direction of motion, however, the speed of the body will increase and hence the work being expended in pushing it will be converted into the increase in the body's kinetic energy.

Work can also be used to change the position of a body in a force field by overcoming the force such as lifting a body from the ground to some given height. The energy or the work done on the body is stored in the form of potential energy by virtue of its position. Once the body begins to fall this potential energy is converted into kinetic energy. Summing up our discussion of work, we see that work can generate the different forms of energy we have so far encountered namely heat, motion (or kinetic energy) and potential energy, and that these different forms of energy are interchangeable.

In addition to the forms of energy that we have so far discussed, there are other forms, which also deserve mention. For example, there is chemical energy, the energy that is released by chemical reactions. It is the source of energy that generates the heat within our bodies and the work and motion we are able to generate with our bodies. There is also light energy, the energy of electromagnetic radiation, which is the form of energy by which the Sun transmits its energy to Earth. Sound is another form of energy, which in fact is the kinetic energy of the air molecules, the oscillations of which create sound.

There is also nuclear energy, the energy generated by the nuclear processes of fission, the splitting of a heavy nucleus of an atom into smaller nuclei and neutrons, and nuclear fusion, where smaller nuclei combine to form a larger nucleus. In both fission and fusion mass is destroyed and converted in energy according to Einstein's famous formula $E = mc^2$. This energy is popularly referred to as atomic energy, when in fact it is nuclear energy. It refers to the energy released both peacefully by atomic reactors and violently by A-bombs and H-bombs. It is also the source of the tremendous amount of energy generated by the Sun. Energy takes many forms. The processes of nature may be considered as the conversion of energy from one form to another.

The Sun is the original source for the different forms of energy we find on Earth with the exception of geothermal energy due to radio-activity. The energy we derive from our food whether animal or vegetable originates with the vegetation of the Earth. The energy stored in plants is a result of the process of photosynthesis whereby plants transform the energy of sunlight into chemical energy and store that energy in the form of hydrocarbons. The energy generated in the Sun by nuclear fusion is converted into heat or internal motion of atoms of the Sun. The internal motion of the atoms causes them to radiate light, which propagates to the Earth and through photosynthesis is converted into chemical energy or food. This chemical energy or food is then converted through oxidation into body heat and motion. The energy of our bodies, therefore, had its genesis in the Sun and has assumed many different forms in order to be transferred to us. It began as nuclear energy and then through successive processes became heat energy, light energy, chemical energy and finally the heat energy and kinetic energy of our bodies.

## Global Warming and Greenhouse Gases

The sunlight from Sun warms the Earth constantly heating the atmosphere and the surface of the Earth. As the atmosphere and the surface of the Earth heat up a certain amount of that heat is radiated back into outer space otherwise our planet would overheat to the point that life on our planet would not be sustainable. There has been a delicate balance in nature in which the amount of incoming energy and outgoing energy are approximately equal which keeps the temperature on Earth in the moderate range. Over the course of time there have been moderate fluctuations in the amount of energy coming from the Sun, due to volcanic activity, fluctuations in the Sun's output and fluctuations in the Earth's orbit about the Sun known as Milankovich cycles. These fluctuations have resulted in Ice Ages, the last one of which ended 10,000 years ago. Another factor affecting the warming and cooling of the planet is fluctuations in albedo, the reflection of sunlight back into outer space due to ice and snow cover. One of the factors affecting the amount of heat radiated out from the Earth into outer space is the amount of carbon dioxide, $CO_2$, in the atmosphere, which is approximately 0.0383% of all the gases in the atmosphere. $CO_2$, however, absorbs a significant percentage of the heat radiated off the surface of the Earth and therefore contributes to the warming of the planet.

The balance of $CO_2$ in the atmosphere before the massive intervention of industrial age human activity was maintained by plants absorbing or breathing in $CO_2$ and exhaling oxygen. The replenishing of $CO_2$ into the atmosphere comes form the respiration of animals breathing in oxygen and exhaling $CO_2$. There is also a certain amount of $CO_2$ that enters the atmosphere due to forest and brush fires caused by lightning and certain amounts of emissions due to volcanic activity. That balance has been seriously eroded by our burning of fossil fuels, which is dumping large quantities of $CO_2$ into the atmosphere. As the amount of $CO_2$ has increased a greenhouse effect has taken place where more energy in the form of sunlight has entered the Earth's atmosphere than is now exiting. There is an analogy with a glass greenhouse because glass is transparent to visible light but is opaque or blocks the lower frequency infrared radiation. This is why a greenhouse is many degrees warmer than the outside air temperature even in the winter.

With the increase of $CO_2$ emissions into the atmosphere due to human activity the overwhelming majority of scientists, especially climatologist, believe that the current round of global warming is due to the greenhouse effect. Ice core data over the past 800,000 years have revealed that the amount of $CO_2$ has varied from values as low as 180 parts per million (ppm) to the pre-industrial level of 270 ppm. Recent measurements of carbon dioxide reveal that concentrations have increased approximately from 313 ppm in 1960 to 383 ppm in 2009. This is why it is believed that the dramatic increase in carbon dioxide levels is responsible for global warming. It is why we now measure our carbon footprint and are trying to reduce our emission of $CO_2$ into the atmosphere. As the planet warms and more ice melts the albedo effects is reduced and the $CO_2$ and methane gases frozen in the ice are released. Not only that but as the oceans heat up they release their $CO_2$ content. Also as the surface area of open water increases there will be more water vapour in the air and water vapour is another greenhouse gas as it is a good absorber of infrared radiation. These effects will only accelerate global warming, which is why we have to take this challenge seriously. There is the danger of a run away greenhouse effect whereby all the ice caps melt, the temperature of the oceans increase dramatically and the warming of the planet accelerates to the point of no return.

## Chapter 10

# Thermodynamics and the Atomic and Molecular Structure of Matter

Both chemists and physicists have contributed to our understanding of the two interrelated subjects we will study in this chapter, namely thermodynamics and the atomic and molecular structure of matter. The word thermodynamics in ancient Greek means literally the movement of heat. This subject is intimately related to an understanding of the gaseous, liquid and solid phases of matter and the transitions between them in terms of the atoms and molecules of which they are composed.

Let us therefore begin our discussion by considering a solid, which as we read earlier in Chapter 8 is not quite solid since the atoms of which it is composed are mostly empty space. The spacing between atoms or molecules if we are dealing with a compound is comparable with the size of the atoms or the molecules themselves. The reason that a solid seems solid is that there exists a binding force between each molecule in the solid. The force arises from the uneven distribution of charge within each molecule or atom. The positively charged nucleus of one atom attracts the electrons of the other atom and vice-versa. But there are repulsive forces as well between all the positively charged nuclei and also between all the negatively charged electrons. On the average, the forces due to these charges cancel each other but a slight residue remains, which accounts for the molecular forces felt between the individual molecules and atoms of matter. The sign of the force depends on the distance between the molecules. There is an equilibrium position where all the electric forces cancel but if the molecules are separated by a distance greater than the equilibrium distance then the force between them is attractive. If on the other hand, the molecules are closer than the equilibrium position then the force is repulsive. It is the repulsive aspect of the molecular force, which prevents us from compressing solids into extremely small spaces. If you wish to experience this force then close this book and try to compress it. After squeezing out the air between the pages, you will find little if any give. You are pushing the paper molecules against one another and they are resisting the force you are creating. The attractive force between the molecules can be felt by trying to tear a solid apart. This force is not always overwhelming as can be demonstrated by tearing a piece of paper in half.

Although there exists an equilibrium position between molecules where the molecular force vanishes, the molecules of a solid do not sit placidly in this position. In fact, the molecules are oscillating wildly about their equilibrium positions. It is this motion, which we identify as heat. It is the kinetic energy of this internal motion of the molecules, which explains why it takes energy to transfer heat to matter. The temperature of a solid represents the rate at which the molecules are moving. The heat of a solid represents the total amount of internal energy due to its internal molecular motion. As the temperature of a solid drops, the molecular motion slows down and hence the amount of heat contained in the solid diminishes. The lowest temperature possible is the point at which all molecular motion ceases. This point known as absolute zero (0 Kelvin) although impossible to achieve exactly, occurs at $-273.15°$ Centigrade.

As the temperature of a solid increases the molecular agitation becomes greater and greater. As the temperature reaches the melting point, which for water is $0°$ Centigrade, the molecular motion becomes so violent that the molecular forces can no longer maintain the structure necessary for the substance to remain a solid.

The substance at this point passes into the liquid phase in which a large number of the molecular bonds have been broken. A number of molecular bonds still remain, however, in the strings of molecules that form the structure of a liquid. These strings of molecules easily pass by each other, which explains the fluid nature of liquids. This ability to flow is also possessed by granulated forms of solids such as sand and salt, and provides a macroscopic picture of the fluid nature of liquids.

The electric forces between molecules in a liquid manifest themselves in a number of ways. The short-range repulsive interaction is responsible for the incompressibility of liquids in much the same manner as it is with solids. The attractive aspect of the molecular interaction on the other hand is responsible for the property of liquids to coalesce. This property is observed as the tendency of water to form itself in drops. At the surface of a drop or any other aggregate of a liquid one observes a surface tension, which arises from the attractive aspect of the molecular forces. The strength of this surface tension is considerable. it enables certain water bugs to walk upon the surface of the water. Although they are heavy enough to sink they are not heavy enough to break the surface tension and they therefore actually walk upon the surface of the water.

Two other forces related to each other and associated with liquids are the forces due to water pressure and buoyancy. The water pressure at any given point is related to the weight of the water between the point in question and the surface, and hence depends solely on the distance to the surface. The distance to the bottom is irrelevant. The pressure pushes equally in all directions. It is the difference in the water pressure at the top and bottom of an object, which accounts for its buoyancy in water.

Let us consider some material, which has the form of a cube each of whose sides is one meter long. We place our cube under water such that the top of the cube is one meter from the surface and the bottom of the cube is two meters from the surface. There are now three distinct forces in the vertical direction acting on the cube. Two of these forces are due to water pressure at the upper and lower surface. The one at the upper surface is pushing down and is less than the one at the lower surface pushing up. The difference of these two forces is precisely equal to the weight of the water displaced by the cube. The third force acting on the cube is the gravitation pull of the Earth and is of course equal to its weight. If the weight of our material is greater than that of the displaced water is will sink and if it is less then it will float. We therefore expect those substances whose density is less than water to float and those whose density is greater to sink.

Archimedes was the first to explain these ideas over 2000 years ago. He made this discovery when he stepped into his full bathtub and observed the water he displaced flow over the sides of the tub. Reportedly, he jumped out of the tub and ran home naked crying "Eureka!" for he realized he could solve the problem of whether the king's crown was pure gold or a fraudulent mixture of gold and some base metal. By comparing the amount of water displaced by the king's crown and an equal amount of pure gold, he could determine if the crown was also pure gold. This episode with Archimedes is one of those rare moments in Greek science when empiricism played a role. It was not quite a planned experiment but an accidental observation. It is, however, one of the times that Greek science was absolutely correct unlike most of the physics of Aristotle that had to be undone.

As with solids the molecules of liquids are in thermal motion. As the temperature of the liquid is increased, this thermal motion increases until the boiling point is reached (which for water is 100° Centigrade). The thermal motion is then so great that the remaining molecular bonds of the liquid phase are broken and the substance enters the gas phase in which there are no bonds between the individual molecules.

Even before the boiling point is reached, however, a certain amount of evaporation of a liquid into a gas takes place. This phenomenon of evaporation is responsible for the creation of clouds over large water masses and hence subsequent precipitation that follows. Evaporation can be explained easily with our molecular model of liquids. At the temperatures of the liquid phase the thermal motion of the molecules is quite violent. Occasionally a molecule near the surface has enough kinetic energy to break its ties with the other molecules and it escapes from the liquid. It then enters the atmosphere as water vapor. At any given temperature the vapor pressure that is the amount of water vapor in the air is fixed. An equilibrium between the number of water molecules escaping the liquid and the number accidentally falling back and being reabsorbed is established.

In the gas phase all of the molecular bonds have been broken. Except for collisions the molecules of a gas are completely independent of each other. Most of the volume occupied by the gas is empty space. A volume of water vapor is about 1700 times that of an equivalent weight of water in the liquid phase.

A gas is characterized by three quantities; its volume, pressure and temperature. The volume of the gas is defined as the volume of the container holding the gas. The pressure is the force per unit area exerted on the sides of the container by the molecules bouncing off the sides of the container. Conservation of momentum demands that each time a molecule bounces off the wall, it transfers an amount of momentum equal to twice its momentum in the direction perpendicular to the wall. The temperature of the gas is a measure of the motion of the individual molecules. It is in fact, as we will show below, directly proportional to the kinetic energy of the molecules.

The original work on gases by Robert Boyle revealed that for a fixed temperature the pressure and the volume are inversely proportional or putting it in an equivalent form, the product of the pressure times the volume is a constant at a fixed temperature. This result is consistent with the atomic picture of a gas since decreasing (increasing) the volume of the gas increases (decreases) the number of molecules colliding with the side and hence increases (decreases) the pressure. Subsequent experimentation revealed that the product of the volume times the pressure is directly proportional to the temperature. By considering the gas at a fixed volume, we find that the pressure is proportional to the temperature. If we extrapolate all the graphs of pressure versus temperature at fixed volume to lower temperatures we find that the pressure is zero when the temperature is –273.15° Centigrade. This temperature represents absolute zero where all molecular motion ceases and hence the pressure is zero. If we convert to the Kelvin scale of temperature where –273.15° Centigrade is 0 Kelvin (and 0° Centigrade is 273.15 K) then for a fixed volume the pressure is directly proportional to the temperature and approaches zero as the temperature also approaches zero Kelvin.

The pressure of a gas at a fixed volume depends on the amount of momentum transferred to the wall per unit time. The momentum transferred per collision as mentioned above is proportional to the momentum of the molecule. The rate at which the molecules strike the wall on the other hand, is proportional to their velocity since the faster they are moving the more likely it is they can reach the wall from any given position in the gas. The pressure is therefore the product of the molecule's momentum times its velocity, which is just twice the kinetic energy of the molecule and therefore the temperature is proportional to the kinetic energy of the molecules.

The result explains why for a given substance that it takes a fixed amount of energy to raise the temperature a fixed amount over a large range of temperatures. The heat of a substance as predicted earlier by Bacon is just the internal kinetic energy of the molecular or atomic motion of the substance. The result that we derived for gases applies as well to liquids and solids.

The molecular structure of matter also helps us to understand the conduction of heat by solids. When two solids of different temperature are in contact with each other the molecules of the hotter body are moving faster than those of the cooler. As the faster moving molecules collide with the slower moving ones they transfer some of their kinetic energy to these molecules. This process goes on until eventually the two bodies are at the same temperature. Convection is the process whereby the transfer of heat takes place through the flow of a warm gas or liquid. It is similar to conduction in that the molecules of the warmer body transfer their kinetic energy or hear through collisions with the molecules of the converting fluid, which in turn transfer their kinetic energy or heat through molecular collisions to the cooler body. The third common process of heat transfer referred to as radiation involves the emission of electromagnetic radiation in the infrared region of the spectrum by the molecules of the hotter of the two bodies as a result of their thermal motion and the absorption of this radiation by the molecules of the cooler body.

In all processes involving the transfer of heat from one body to another it is always found that the total energy is conserved. This is the first law of thermodynamics. In terms of the atomic description of matter, the conservation of energy is a result of the fact that the molecular collisions conserve kinetic energy. In processes such as friction where mechanical energy is converted into heat energy conservation of energy holds for the same reason. In this process the average motion of all the individual molecules is reduced and the energy that is lost by reduction of the overall kinetic energy of the body is completely randomized into random thermal motion. Thus the process whereby mechanical energy is converted into thermal energy conserves energy in accordance with the first law of thermodynamics, which is a statement of the conservation of energy.

Friction also increases the disorder of the universe in accordance with the second law of thermodynamics, which states that any spontaneous change in a physics system increases its entropy or disorder. Entropy is a quantitative measure of the disorder of a system whose exact definition need not concern us here. We will use the term synonymously with disorder. The natural tendency of the entropy or disorder of physical systems to increase spontaneously results from the fact that states, in which the order increases are highly unlikely to occur from a probabilistic point of view. The probabilities become exceedingly small because of the enormous number of molecules composing any given system. The best way of illustrating this concept is to consider a pile of papers place outdoors on a windy day. As soon as the winds gust, the orderly pile of paper will become a random collection of papers scattered all about. If, on the other hand, I started with a random scattering of paper I could not expect the wind to blow the papers into a nice neat pile. This illustrates the natural tendency of physical processes to increase the disorder or entropy of the universe.

Another example of the natural increase entropy can be seen by considering a box with two partitions and a wall between them. In the left hand side of the box we place 2 grams of hydrogen gas or $6 \times 10^{23}$ hydrogen molecules, while we leave the other side empty. Once we remove the wall between the two partitions there will soon be approximately $3 \times 10^{23}$ hydrogen molecules on each side of the box. Although it is physically possible for all of the molecules to move back to one side of the box the probability of this is so small, 1 part $8^{100,000,000,000,000,000,000,000}$ that it can be effectively ruled out. Life would be fairly dangerous if this were possible; imagine choking to death as all the air molecules on your side of the room suddenly moved over to the other side of the room.

An increase of entropy occurs whenever two bodies of different temperature are placed in thermal contact. Invariably the warmer body loses heat to the cooler until the two of them are at the same temperature. Although conservation of energy does not forbid two bodies of the same temperature transferring heat between them so that one becomes warmer and the other cooler, this process does not occur naturally. When two bodies are at different temperatures more of the molecular motion is with the warmer body. When these two bodies are placed in thermal contact so that motion becomes more evenly distributed between the two bodies, resulting in a more random distribution of the motion. If one wants to distribute heat from one body to another when both are initially at the same temperature one needs a heat pump, a refrigerator, or an air conditioner and the energy it takes to run these devices. So it takes work to create a temperature difference. On the other hand if one has a temperature difference between two gases there will be a flow of molecules between them, which can be harnessed to do work, which is basically how a steam engine or a gasoline internal combustion engines works.

# Chapter 11

# Electricity and Magnetism

A Newtonian description of nature requires an understanding of the basic forces that operate between the various constituents of the universe. So far, the only force we have discussed in any detail is the gravitational force. We shall now turn our attention to the electric and magnetic forces, which, we will soon discover, are intimately connected to each other and are referred to collectively as the electromagnetic interaction. The electromagnetic interaction plays a major role in determining both the structure of matter and the general interactions of matter as we discussed in the last chapter. The reason for this is, that of the three major constituents of gross matter, the proton, the neutron and the electron, two of them, the electron and proton, are charged and hence, exert an electric force whereas all three possess magnetic moments and hence, exert magnetic forces. As a result of this the internal structure of atoms and molecules, as our future studies will reveal, are governed by the electromagnetic interaction. For example, the negatively charged electrons of an atom are held in their orbits about the positively charged nucleus by the electric force. Not only are the forces inside the atom and molecule electromagnetic, but also the forces between atoms and molecules are electromagnetic. Molecular bonds are responsible for the structure of gross matter. Hence, the resistance one feels when one tries to penetrate solid matter is electromagnetic as is the force exerted by a coiled spring. The forces produced by chemical action are also electromagnetic since all chemical reactions are governed by the electric properties of an atom's outer electrons. The production of light, as we will soon see, is also a result of the electromagnetic interaction.

In addition to the natural phenomena referred to, we encounter the effects of the electromagnetic interaction in a steadily increasing number of devices and machines invented by humans for their convenience and pleasure such as the light bulb, the neon sign, the radio, the television, the computer and the laser, not to mention the electric streetcar, the electric stove, the electric dishwasher, the electric iron, the electric vacuum cleaner, the electric can opener, etc. etc. etc.

Let us begin our study of the electromagnetic interaction by turning our attention to the electric force, which, in some ways, is like the gravitational force. The strength of the electric force, like the gravitational force, is inversely proportional to the square of the distance between the two interacting bodies. The magnitude of the strength is proportional to the product of the two charges. The force is represented by the equation $F = kq_1q_2/R^2$. In addition to the inverse square law the force in both cases acts along the line joining the two bodies and is equal and opposite for the two bodies as is illustrated in Fig. 11.1.

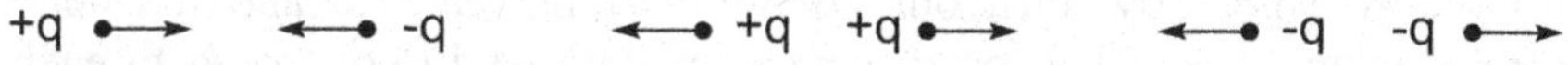

+q •→     ←• -q       ←• +q   +q •→       ←• -q   -q •→

Fig. 11.1

The electric force differs from the gravitational force, however, in two vital ways. First of all, the gravitational force is only attractive whereas the electric force can be both attractive and repulsive depending on the signs of the charges involved. A charged particle is either positive like a proton or negative like an electron. The proton and electron along with the neutron are the basic elementary particles that make up the atom. The proton and electron have equal and opposite charges +e and –e; and the neutron is electrically neutral. All of the charges that are found in nature aside from those of exotic elementary particles created in physics labs are due to electrons and protons. Charges of the same sign repel each other whereas charges with opposite signs attract each other. The other difference in the two forces is the fact that the strength of the electric force is considerably greater than that of the gravitational force. In fact, the electric attraction of a proton and an electron is $10^{40}$ times stronger than their gravitational attraction. This accounts for the very strong forces, which hold the atom together, create the chemical bonds in molecules and produces the molecular bonds in solids and liquids.

Because of the fact that the electron and proton are exactly equal and opposite in charge, gross matter is electrically neutral. In fact, if there was a slight difference in the magnitudes of the electron's and proton's charge of only one part in $10^{20}$ macroscopic matter would be completely unstable for the repulsive forces generated by such a minute difference in charge would be enough to completely destroy all the molecular bonds, which hold matter together and scatter it to the four corners of the universe.

Fortunately for our existence the equality of the magnitude of the charges for the proton and electron, two particles, which differ in so many other ways, is apparently identical. Macroscopic matter becomes electrically charged, however, whenever there is a slight excess or shortage of electrons. This occurs, for example, when a rubber rod is stroked by a piece of cat fur in which case electrons are transferred by friction from the fur to the rubber rod. The rubber rod has an excess of electrons and hence, has a net negative charge whereas the fur contains more protons than electrons and hence, has a net positive charge. If, after rubbing the rod with the fur, one were to put these two objects in contact, electrons attracted by the positive charge of the fur would flow from the rod to the fur until electric neutrality was once again established. Electric neutrality can also be established by placing a copper wire between the rod and the fur, which would permit the passage of an electric current of electrons to flow between them.

Not all materials, however, permit the passage of an electric current. Some materials such as wood, asbestos and rubber, referred to as insulators, do not permit the flow of an electric current because all of their electrons are tightly bound by the chemical bonds holding these materials together. In certain materials, however, such as metals, not all of the electrons are so tightly bound in their atoms. These materials, referred to as conductors, permit the flow of an electric current. When an electric current flows, electrons do not flow from one end of the wire to the other. The electrons in a conductor behave more or less like the molecules in a gas. When there is no current flowing in the conductor, the free electrons move back and forth within the wire in a random fashion colliding with each other and the atoms making up the wire. When a current is flowing there is a general drift of the electron in a particular direction. The net effect is that a current flows from one end of the wire to the other although no actual electron makes this trip. The wire is actually electrically neutral with as many electrons flowing into any one segment as flow out of that segment. If more electrons flow in one direction than the other then there is a flow of current in the opposite direction because the charge on an electron is negative and the flow of current is defined as the direction of the flow of positive charge as is illustrated in Fig. 11.2. The heat generated by an electric current is due to the collisions of the electrons with the atoms making up the conductor. These collisions are also responsible for the resistance of the conductor. This explains why, for a given current, the amount of heat generated in a conductor is proportional to its resistance.

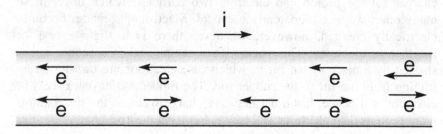

Fig. 11.2

Rubbing a rubber rod with a piece of cat's fur and placing a wire between them is not the most efficient way of creating an electric current. However, the more efficient methods of producing an electric current are based on the same idea of creating two polarities of charge, which a flow of current will neutralize. This is the principle of a battery with the difference that the separation of the positive and negative charge is a perpetual process brought about by a chemical reaction. A battery can be created simply by placing two different metal rods in an acid solution. When the acid acts upon the metals, it combines chemically with the metal depositing an excess of electrons on the rods. Since the acid works on one of the metal rods of the battery faster than the other, more electrons build on one of the metal rods faster than on the other rod. If a wire is placed between the two metal rods of the battery an electric current will flow in order to neutralize the excess of electrons. However, this flow of electrons activates the battery to continue acting chemically on the metal rods to produce an excess of charge, which causes a current flow and so on until the acid finally eats away one of the rods and then the battery can no longer generate an electric current.

From the study of electric currents generated by devices like the battery, it was discovered that electric currents exert a force on each other. This force cannot be attributed to the electric force. Although an electric current involves the flow of electrons, the current in a wire is actually neutral. As stated earlier, the current is due to the drift of the free electron in the copper wire in one particular direction. This direction is actually opposite to the flow of the current. This apparent backward way of defining the current is due to the historical accident that Benjamin Franklin defined a current as a flow of positive charge at least one hundred years before the electron was actually discovered. Although there is a current due to the drift of the electrons, the wire itself is electrically neutral. If one examines a given section of a wire, as much charge flows out of it as into it. The force observed between wires carrying currents is therefore not due to the electric force but some other force.

It was also discovered that an electric current exerts a force on a magnetic compass. In fact, one can make an artificial magnet by wrapping copper wire around an iron bar and passing a current through the wire. It is clear from these two experiments that an electric current behaves like a magnet and that the force between two electric currents is magnetic. Further study showed that the magnetic force, like the electric force, is also inversely proportional to the square of the distance between the two currents. Since the current in a wire has a net charge of zero, one cannot ascribe the magnetic force solely to the charge of the particles within. It is clear that it is the motion of the charge particles that produces this new force, which differs from the electric force in a number of ways.

Perhaps the best way of comparing the two forces is to consider the interaction of two positively charged particles moving parallel to each other. There is a repulsive electric force between these two charges whose strength is proportional to the product of their charges divided by the square of the distance between. In addition to this electric force there is also an attractive magnetic force whose strength is also proportional to the product of the charges divided by the square of the distance between them. The magnetic force is also proportional, however, to the product of their velocities divided by the velocity of light squared. Since the velocity of a particle can never be greater than the velocity of the light, the magnetic force is always less than the electric force.

The force depends on the relative direction of the two currents in a complicated fashion. When the two currents are parallel (anti-parallel) the force is attractive (repulsive). The force acts along the line connecting the two currents and is equal and opposite. For other configurations, the force is not always directed along the line connecting the two currents and it is not always equal and opposite.

The connection between the magnetic properties of an electric current and a lodestone or magnet is easily made by considering the atomic structure of a magnet. All atoms, because of their electrons orbiting the nucleus, have equivalent electric currents, which can exert magnetic forces. Since the orientation of atoms in most matter is so completely random, the effects of each individual atom's magnetism cancel. In certain very select materials, such as lodestones, the atoms are oriented in such a way that the magnetic forces exerted by individual atoms can add up constructively to create a rather strong magnetic force.

This explains why a magnet loses its magnetism if it is dropped or heated since, in both of these cases, the special orientation of the magnet's atoms are destroyed. This also explains why the north pole of one magnet attracts the south pole of another since it is in this position that the internal currents of the atoms are parallel and hence, attractive. When one of the magnets is rotated so that now two north poles are facing each other or two south poles, then the internal currents are anti-parallel and the two magnets repel each other.

Both the electric and magnetic interactions of charged particles discussed above have a magical quality about them in the sense that the charged particles interact with each other at a distance without any apparent physical connection between them. With the exception of the gravitational force, which also has this magical property of action at a distance, all the other forces between bodies require some kind of physics contact. The concept of action at a distance is very difficult to comprehend. Try conceiving of how you personally could move some object without coming into physical contact with it. This is the work of a magician. Yet every proton, every electron, is a magician since they exert forces on each other through a vacuum with absolutely nothing between them. How can one account for this?

Michael Faraday invented the concept of the electric and magnetic field in an attempt to understand this mystery. According to his idea, each charged particle created an electric field about it. If the particle is in motion, then, in addition to the electric field, it also creates a magnetic field. Both the electric and magnetic fields spread throughout all space. The field, at any given point in space, is inversely proportional to the square of the distance from the charged particle generating the field. A charged particle finding itself in the electric field generated by another charged particle will experience an electric force according to the strength of the electric field and its own charge. If this particle is in motion then it will experience a magnetic force proportional to its charge, its velocity and the strength of the magnetic field.

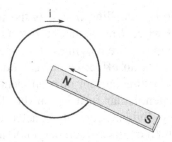

Fig. 11.3

While the concept of the electric and magnetic field might solve the mystery of action at a distance, it leaves us with a new mystery; to wit, how does a charged particle create an electric or magnetic field at some distance from it without a connecting medium. Faraday and his successor actually believed in the existence of an invisible medium, which they called aether, which, according to them, filled up all of space. This concept survived until the advent of the theory of relativity when Einstein showed that the existence of an aether was inconsistent with experimental facts. The electric and magnetic field, as far as I am concerned, do not really exist in the same sense that charged particles exist. The field concept is an abstraction, a pictorial description of the electric and magnetic forces. It is, nevertheless, a conceptual construction, which is very useful. The concept of electric and magnetic fields will help us to understand the phenomenon of electrical induction. It will also help to explain the creation, absorption and propagation of light as an electromagnetic phenomenon.

The phenomenon of electrical induction was first discovered by Faraday, who observed that an electric current momentarily exists within a loop of wire into which a magnet has been thrust as is illustrated in Fig. 11.3. The current flows only while the magnet is in motion either being inserted into the loop or being withdrawn from it. It was not just the presence of the magnet but its motion that induced an electric current. Faraday discovered that a current could also be induced by moving the loop of wire with respect to the magnet. In other words, it is the relative motion of the magnet and the loop, which induces a current. If the current flows in a clockwise direction when the magnet is inserted, then it flows counter-clockwise when the magnet is removed. Furthermore, if the opposite pole of the magnet is inserted into the loop, then the current also changes direction. An understanding of the mechanism of electric induction can be made by recalling that it is the relative motion of the magnet and the loop of wire, which causes a current to flow.

Let us consider a loop of wire moving with respect to the magnet. The motion of the loop creates an effective current out of every electron in the wire. The direction of this current of each electron in the wire is not along the wire but perpendicular to it i.e. in the direction of the wire's motion. Each of these individual currents is acted upon by the magnetic field of the magnet, which exerts a force perpendicular to the direction of motion of the wire and hence creates an effective electric field. It is this force, which causes the induced current to flow in the looped wire. The same effect occurs when the magnet is moving and the wire is at rest since it is only the relative motion of the two, which matters.

Electrical induction also operates with two loops of wire facing each other. If an electric current flowing in the first loop changes in any way, it causes a momentary current to flow in the second loop. The principle is the same as that of electric induction caused by a moving magnet. When the current in the first loop of wire changes, the magnetic field generated by this current changes at the position of the second loop. In the case of electric induction involving the relative motion of the loop of wire and the magnet that we just discussed, the magnetic field at the loop of wire is also changing when a magnet is thrust into the loop of wire. It is clear from these two cases that the induced electrical current is caused by a changing magnetic field. Since the induced electrical current is caused by an induced electric field, the phenomenon of induction can be expressed totally in terms of fields. A changing magnetic field induces an electric field, which in turn causes the induced electric current to flow. Faraday's law states that the strength of the induced electrical field produced is proportional to the rate of change of the magnetic field and acts perpendicular to the direction in which the magnetic field is changing.

# Chapter 12

# Electromagnetic Radiation and
# Wave Behaviour

Faraday's concept of the electric and magnetic field was a great aid to his experimental work. Faraday's field concept, however, was absolutely crucial to the mathematical and theoretical work of James Clerk Maxwell. Maxwell was able to express all the laws of electricity and magnetism in terms of four very simple equations, which relate the electric and magnetic fields and show their intimate connection. In fact, from the symmetry of his equations, he predicted that, in analogy to Faraday's law of electric induction in which a changing magnetic field creates an electric field, that a changing electric field would create a magnetic field. This prediction of magnetic induction, i.e. the induction of a magnetic field by a changing electric field was immediately confirmed by experimental work and verified the validity of Maxwell's equations.

Maxwell was able to obtain still a more significant insight into electromagnetic phenomenon from the study of his equations. He discovered the existence of a solution to his equations in which there is an absence of charge and in which the electric and magnetic fields behave like a wave. He associated this solution with the phenomenon of light, which he recognized as electromagnetic radiation. As a result of this insight, he was able to explain the emission, absorption and propagation of light.

The concept of an electromagnetic wave is easy to understand once its relation to electric and magnetic induction is realized. Consider an electric field oscillating at some point in space. Then, by magnetic induction, (or since a changing electric field produces a magnetic field) the oscillating electric field creates a magnetic field perpendicular to itself in its immediate neighborhood. This oscillating magnetic field, as a result of electric induction, then creates an oscillating electric field, which, in turn, induces an oscillating magnetic field, which, in turn, induces an oscillating electric field and so on and so forth. In this way, an electromagnetic wave propagates through empty space at the velocity of light, denoted by c, which is 300,000 kilometers per second (186,000 miles per second) or $3 \times 10^8$ meters per second.

The production of electromagnetic radiation can be achieved by causing a charged particle to oscillate back and forth since this causes the electric field associated with the charged particle to also oscillate. This is precisely how radio waves, another form of electromagnetic radiation, are produced in radio antennas. A current of electrons is made to oscillate up and down in the antenna at a given frequency in order to broadcast radio waves.

The absorption of electromagnetic radiation occurs as a result of charged particles interacting with the oscillating electric and magnetic fields of the electromagnetic radiation. For example, the eye detects visible light when the electrons in the retina become activated by the electric and magnetic fields of the light ray.

Electromagnetic radiation comes in a variety of different forms such as microwaves, radio waves, infrared (heat) radiation, visible light, ultra-violet radiation, x-rays and gamma rays. All of these forms of electromagnetic radiation are identical in the sense that they are oscillating electric and magnetic fields, which all propagate at the velocity of light. They differ only in that each one represents a different range of frequencies and hence, wavelengths. The frequency, f, of a wave is the number of times per second that the electric and magnetic fields oscillate back and forth. It is inversely proportional to the period, T, of the wave defined as the time for one complete oscillation. The wavelength, $\lambda$, is the distance between successive maximums of the field as shown in Fig. 12.1 below and is proportional to the period and inversely proportional to the frequency. The wavelength, $\lambda = cT = c/f$. A list of the frequency and wavelength of the various forms of electromagnetic radiation is given in the accompanying table.

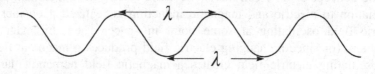

Fig. 12.1

**Table of radiation, wavelength and frequency of electromagnetic radiation.**

| Type of wave | Wavelength, $\lambda$ (cm) | Frequency, $f$ (sec$^{-1}$) |
|---|---|---|
| radio waves | $10^3$ to $10^4$ | $10^8$ to $10^9$ |
| microwaves | $10^{-3}$ to $10$, | $10^9$ to $10^{12}$ |
| infrared (heat) | $10^{-4}$ to $10^{-3}$ | $10^{13}$ to $10^{14}$ |
| visible light | $4$ to $7 \times 10^{-5}$ | $10^{15}$ |
| ultra-violet | $10^{-7}$ to $10^{-6}$ | $10^{16}$ to $10^{17}$ |
| x-rays | $10^{-9}$ to $10^{-7}$ | $10^{17}$ to $10^{19}$ |
| gamma rays | less than $10^{-9}$ | $10^{19}$ and higher |

The range of frequencies of visible light and the range of frequencies that the Sun radiates with maximum intensity exactly overlap. The eye has been biologically adapted through natural selection and evolution so as to detect the electromagnetic radiation of the Sun. If human life had developed on a planet orbiting a star, which emitted principally infrared radiation then, through the process of evolution, one would expect visible light for these people to be in the infrared range. The visual detection of objects using infrared radiation has actually been achieved by the military. They have developed a photographic film sensitive to infrared rays, which they have used for aerial photo reconnaissance at night.

In addition to perceiving electromagnetic radiation, the eye is also sensitive to the different frequencies in the visible light range. This accounts for our colour vision. Each colour corresponds to a different range of frequency in the visible range as is listed in the table below. The order of the colours in the list is exactly the same as the rainbow. This is no coincidence. The Sun radiates visible light of all frequencies and hence, all colours. White light is merely a combination of all the colours. When white light or sunlight propagates through water such as a raindrop or glass such as a prism, the light ray is bent both upon entering and leaving the medium of water or glass. The amount of bending a ray of light experiences depends on the frequency of the ray; the greater the frequency, the more it is bent. The different colours that compose light are therefore separated when they propagate through a medium like water or glass and hence, one observes a rainbow.

The knowledge that sunlight is composed of all the colours helps us to understand why the sky is blue and why sunsets and sunrises are red.

When one looks up at the sky on a cloudless day, one observes blue light. This light is sunlight, which has been absorbed by air molecules in the upper atmosphere and reradiated towards Earth. Because the amount of light reradiated at any one particular frequency is proportional to the fourth power of the frequency, more blue light than red is reradiated towards Earth and hence, the sky appears blue. At sunset or sunrise, light arriving from the Sun has a thicker envelope of air to travel through in order to reach us. Since more blue light than red has been absorbed out of the beam of sunlight as it travels through the atmosphere, the light reaching us during a sunset or sunrise appears red.

| Colour | Wavelength, $\lambda$ (cm) | Frequency, f (sec$^{-1}$) |
|--------|----------------------------|----------------------------|
| red    | $7 \times 10^{-5}$         | $4.2 \times 10^{14}$       |
| orange | $6.5 \times 10^{-5}$       | $4.6 \times 10^{14}$       |
| yellow | $6 \times 10^{-5}$         | $5.0 \times 10^{14}$       |
| green  | $5 \times 10^{-5}$         | $6.0 \times 10^{14}$       |
| blue   | $4.5 \times 10^{-5}$       | $6.7 \times 10^{14}$       |
| indigo | $4.3 \times 10^{-5}$       | $6.9 \times 10^{14}$       |
| violet | $4 \times 10^{-5}$         | $7.5 \times 10^{14}$       |

Maxwell's identification of light with oscillations of the electric and magnetic fields explains the wave nature of light. Long before Maxwell's identification, it was realized that light behaves as a wave. The wave nature of light was first suggested by Christian Huygens, a contemporary of Newton. In fact, he and Newton had a long-standing controversy concerning the nature of light. Newton adopted the position that light was a beam of particles and hence, could not display wave behaviour. Huygens had a difficult time convincing the scientific world of the wave nature of light because of the formidable reputation of his scientific foe. After the results of a number of experiments corroborating Huygens point of view became known, however, the science community finally adopted the wave picture of light.

By an ironic twist of fate, however, experiments performed in the early part of the 20th century have revealed that, although light behaves in many situations as a wave, there are instances when it also behaves like a beam of particles. So, there is also a sense in which Newton was correct. However, from the point of view of the experimental evidence that was available to Huygens and Newton, it was Huygens who made the more accurate interpretation of the data.

We will defer our discussion of the wave-particle duality of light to the time when we discuss atomic physics, and turn our attention instead to the wave nature of light. Let us first consider the nature of wave behaviour in general by discussing a more familiar example, namely, the waves that travel on the ocean. When one looks at the surface of the ocean on a windy day, one sees alternate rows of crests and troughs. As one observes the movement of the water, one observes that the crests and troughs are moving toward the shore. One might erroneously conclude that the water is moving towards the shore but, in fact, the water composing the wave is actually moving up and down. It is oscillating in the direction perpendicular to the direction in which the wave is moving. This can be easily verified by watching a buoy bobbing up and down in the water. At the shore, it is true that at certain moments water moves toward the shore, but it is also true that, at other moments, an equal amount of water moves away from the shore as the wave washes back into the sea.

One should not confuse the two different types of motion one encounters in wave behaviour. One motion is the motion of the wave or really, the waveform, which is continuous and unidirectional. The other motion is the actual movement of the medium, which is always an oscillatory motion. In the example of waves upon the ocean, the oscillatory motion of the medium is perpendicular to the motion of the waveform. This type of wave is called a transverse wave and is differentiated from a longitudinal wave in which the medium oscillates back and forth in the same direction in which the wave moves.

Perhaps the best-known example of a longitudinal wave is a sound wave. A sound wave requires the existence of transmitting medium. Most of the sound waves with which we come in contact propagate through the atmosphere although sound waves can also propagate through solids and liquids. There are no sound waves on the surface of the Moon however, because there is no atmosphere.

Let us consider the production and propagation of sound waves in our atmosphere. A sound wave is produced as a result of the rapid vibration of some object, like a violin string for example, which causes the molecules of air surrounding it to move back and forth like the vibrating string. This causes alternate condensations and rarefactions of the air molecules. As the string moves to the right, it pushes the air molecules together creating a condensation; as it moves back to the other direction, it leaves a rarefaction. The motion of the air molecules back and forth is in the same direction as the motion of the vibrating string. The vibrating column of air adjacent to the string will, in turn, cause the column of air adjacent to it to vibrate and then, this column of air will act on the column adjacent to it and so on and so forth and in this way, the sound wave will propagate through the air. Each column of air will vibrate back and forth and hence, there will be no net displacement of the air as the sound wave propagates from the vibrating string to the ears of a listener.

The air molecules that are contact with violin strings will not come in contact with the listener's ear. The vibration of these air molecules will propagate through the air to the listener's ear as a consequence of the contact the air molecules make with each other through collisions. The vibrations of the final column of air adjacent to the eardrum of listeners will cause their eardrum to vibrate with the same frequency of the original violin string. The vibrations of the eardrum are transmitted through tiny bones to a cavity, the cochlea, containing a fluid where they activate nerve cells, which transmit the information to the brain. The human ear is capable of detecting frequencies in the range from sixteen vibrations per second to twenty thousand. The greater the frequency of the sound wave, the higher the pitch that we detect. The loudness of the sound wave depends on the strength of the vibrations or on the distance through which the string vibrates. The harder the string is struck, the greater is the amplitude of its vibration and hence, the louder the sound. The frequency of the string does not depend on the strength with which it is struck but rather on the length of the string, its thickness and the amount of tension with which it is strung.

We have now considered both transverse waves, (waves on the ocean) and longitudinal waves (sound waves). In both cases, the wave is transmitted as the result of oscillatory motion of a physical medium. In the case of ocean waves, the water was oscillating up and down in the direction transverse to the propagation of the wave. In the case of the sound wave, the air molecules were oscillating back and forth in the direction longitudinal to the wave motion. We encounter a somewhat different situation when we consider electromagnetic radiation since no medium is required to propagate the wave motion. Light can travel in a vacuum. i.e. empty space. Instead of the oscillation of some physical medium, electromagnetic waves involve the oscillation of the electric and magnetic fields. The oscillation of these fields is transverse to the direction of the wave propagation and hence, light is a transverse wave.

We are still left with the mystery of how a wave is able to propagate through empty space. This mystery is related to the mystery of action at a distance discussed earlier in connection with electric and magnetic forces. The solution to these two related mysteries, provided by Faraday and Maxwell, is the concept of a field. Later in this book, after we have studied more about elementary particles and their basic interactions, we shall return to this mysterious question and consider another possible solution.

Although it is more difficult to conceive the wave nature of light than that of the ocean because of the absence of a concrete medium, light, nevertheless, displays exactly the same wave behaviour characteristic of waves on the water. Let us consider two phenomena characteristic of waves, namely, linear super-position and interference. If I drop a stone into a quiet pond of water, a wave front with a circular shape will propagate from the point where the stone enters the water. If two stones are dropped into the water a short distance apart, two circular waveforms will propagate and interfere with each other.

The two waves will flow through each other without affecting each other, that is, after passing each other the two waves are exactly the same as they were before, i.e., they retain their circular form. In the region where they meet, however, they interfere with each other. The motion of the water up and down due to the two waves will either add together or subtract depending on whether or not two crests arrived at the same point or a crest and a trough arrived at the same point. If two crests arrive at the same place, then, the waves add such that a crest is created higher than that of a single wave is created. If, on the other hand, the crest from one wave arrives at the trough of another, then, the two waves can momentarily cancel so that it appears there is no disturbance of the water at all at this point. However, an instant later, as the two waves propagate past each other, one observes the two waves again.

Light also can interfere constructively or destructively with itself just like water waves. Let us consider light from the same source shining through two slits of some opaque material. This situation is analogous to the dropping of two stones since spherical light waves will emanate from each of the two slits. If we now observe the light from these two slits projected on a screen, we will observe a pattern of alternating illuminated and dark patches. Those positions, which are illuminated, are the places where the light from the two sources arrived in phase and the dark points where the two beams of light arrived out of phase. It was the two-slit interference experiment first performed in 1789 by Thomas Young, which finally settled the controversy between Newton and Huygens concerning the wave or particle nature of light.

# Chapter 13

# Prelude to Relativity

With Maxwell's description of light as electromagnetic radiation, the physicists of the nineteenth century felt that they had a complete description of the physical universe. Perhaps there were certain details to be studied but they believed that all of the basics were understood. In actuality, they were on the eve of a revolution in physics that would shake the foundations of their thought as much as the Copernican revolution had shook the thought of their predecessors some three hundred years earlier. Their entire notion of space and time would change as a result of this revolution, which Einstein's Theory of Relativity would bring. But before discussing the nature of this revolution, let us first consider the experimental observations that lead to the theory of relativity.

Maxwell had shown through his equations describing the electric and magnetic fields that light was an electromagnetic wave, which traveled at the velocity $c = 2.998 \times 10^8$ m/sec (or approximately 300,000 kilometers per sec). The first measurement of the velocity of light was made by Roemer in 1676 making use of astronomical observations of the eclipses of the Jovian moons by Jupiter itself. The satellite Io orbits Jupiter every 42 hours. Roemer noticed that the time interval between successive reappearances of Io varied depending on the position of the Earth. When the Earth travels along its orbit about the sun from position A to B the time interval between successive reappearances of Io increases. This is due to the fact that the Earth is farther from Jupiter after each successive reappearance and an extra amount of time is required for the light reflected from the moon to reach the Earth. As the Earth travels back from position B to A, the time intervals for successive reappearances of Io decreases. The overall time lag as the Earth travels from position A to B is 22 minutes, which is the time required for the light to travel the diameter of the Earth's orbit.

Using the value of the diameter of the Earth's orbit about the sun, Roemer calculated the velocity of light, c, to be $2.14 \times 10^8$ m/sec. This value is fairly close to the presently known value of $3 \times 10^8$ m/sec. The chief source of error in his calculation was due to the poor estimate in his day of the distance between the Earth and the sun. Although Roemer's determination of c was not very precise, one cannot help but be impressed by the ingenuity of his approach.

A more accurate determination of c was made in 1725 by James Bradley, who also made use of astronomical observations. He observed that the stars that lie above the plane of the Earth's orbit seem to move about a very tiny circle in the sky in the course of a year. This phenomenon, known as the aberration of starlight, is due to the fact that the angle at which the star is observed changes as the Earth moves about its orbit and hence, the star appears to also move in a circle. The angle at which the star is observed depends on the velocity of the Earth. As illustrated in Fig. 13.1, in order for the starlight coming directly from overhead to enter the telescope and subsequently be observed, the telescope must be tipped by an angle $\alpha = v/c$ where v is the velocity of the Earth. From his knowledge of the Earth's velocity, and his measurement of the angle of aberration, Bradley was able to determine the velocity of light. He found the value $3.1 \times 10^8$ cm/sec, a value extremely close to the modern value.

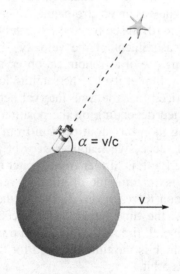

Fig. 13.1

More accurate measurements of c were performed without recourse to astronomical observations using terrestrial-based apparatus, which utilize either a rotating mirror or else a rotating cogwheel. A very accurate measurement was made by Michelson using a rotating mirror involved an optical path of 22 miles between Mt. Wilson and Mt. San Antonio in California in 1880.

One of the triumphs of Maxwell's equations was that the velocity of the electromagnetic waves that he predicted was in agreement with the measured value of the velocity of light. Maxwell's identification of light as electromagnetic radiation explained a great deal about the nature of light. One mystery still remained, however, namely, how is it that the light wave can propagate through empty space? It was impossible for physicists to conceive of a wave without some medium in which it could travel. One need only consider the wave motion of water or sound waves to see that these waves could not exist without their medium. Using these analogies, Maxwell postulated the existence of a transmitting medium for light, which he called the luminiferous aether. The properties of the aether had to be somewhat unusual in order for it to serve the role for which it was invented. It had to fill all space uniformly since the velocity of light is the same throughout space. It also had to penetrate substances such as air, water and glass since light propagates through these materials. Furthermore, this transparent substance could not, in any way, interfere with the motion of corporeal bodies such as the planets since no evidence for the resistance to the motion of corporeal bodies in empty space could be found. In fact, the law of inertia clearly states that the velocity of a body will remain constant as long as no force is acting upon it. The aether can therefore only act on matter through their electric charge. The aether was purported to be the medium through which light propagates and through which electric and magnetic forces between charged particles are transmitted.

The experimental detection of the aether became a challenge to the imagination of the nineteenth century scientists, in particular, Albert Michelson, the man who had so accurately measured the speed of light. Michelson argued that since the Earth is in motion about the sun it must be moving through the aether. Therefore, a beam of light propagating in the direction of the Earth's motion through the aether would have the velocity $c - v$ relative to the Earth. The Earth moves through the aether with the velocity v and the light moves through the aether in the same direction with the velocity c and therefore, their relative velocity is $c - v$.

Light propagating in the opposite direction to the Earth's motion would have the relative velocity c+v. On the other hand, light traveling in the direction perpendicular to the Earth's velocity would have a relative velocity of $c\sqrt{1-v^2/c^2}$ .

Michelson set out to measure the differences in these relative velocities with the help of a physicist named Morley. In 1887, they designed an apparatus, called the Michelson–Morley interferometer illustrated in Fig. 13.2, which could detect the differences of these velocities. Using a series of mirrors, one of which was half-silvered, light from a single source was divided into two paths and then brought back together again. The light travels along identical paths except for one segment in which the light traveled back and forth perpendicular to the Earth's velocity on one path and parallel to the Earth's velocity on the other path. Because the relative velocity of the light is different along these two paths, as outlined above, it takes longer for the light to travel along the path in which it moves only parallel to the Earth's motion. Because of this time delay, one expects to the two beams of light to interfere with each other. The arrangement of light source, mirror and telescopic detector were all mounted on a huge piece of sandstone, which was floating in mercury. This enabled the apparatus to be easily rotated so as to align one of its axes with the Earth's velocity. After carefully and slowly rotating their interferometer, Michelson and Morley were unable to measure any interference.

The inability of Michelson and Morley to observe the effects of the aether was very disturbing to the physicists of their time. Their experiment was repeated and refined a number of times, always with the same result. One of the preoccupations of physicists during this period was an attempt to explain away the results of the Michelson–Morley experiments.

One of the first attempts in this direction was to claim that the aether was dragged along by the Earth's motion and hence, the detection of the relative motion of the aether and the Earth was impossible. This explanation was easily dismissed, however, since the phenomenon of aberration of starlight would not have been detected if the aether were dragged by the Earth. One observes a star at an angle $\alpha = v/c$ because of the velocity v of the Earth perpendicular to the direction in which the starlight is propagating. If the aether was dragged along with the Earth, then, the starlight would also move with the velocity v in the direction of the Earth's motion and hence, it would no longer be necessary to tilt the telescope through the angle of aberration to capture the starlight.

MICHELSON–MORLEY APPARATUS

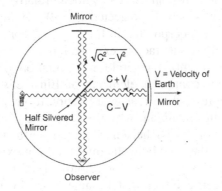

Fig. 13.2

The Irish physicist, Fitzgerald, proposed that the negative result of the Michelson–Morley experiment could be explained if one assumes that matter in motion contracts in the direction of its motion through the aether. He claimed that the contraction, which increases as the velocity of the matter increases, was due to the pressure of the aether wind encountered as the body moved through the static aether. If a body has a length, $L_o$, at rest then, as a result of its motion, it would have the length $L_o \sqrt{1 - v^2/c^2}$. Unfortunately, this contraction could not be measured since any meter stick employed to measure this contraction would also contract. This reduces Fitzgerald's hypothesis to an ad hoc (after the fact) status. His idea is not without merit, however. For one thing, Fitzgerald, using his formula, correctly predicted that no material body could ever travel faster than the speed of light, c, since at this speed the length of the body, $L = L_o \sqrt{1 - v^2/c^2}$ goes to zero. Secondly, the contraction, which he attributed to an actual physical force reappeared later in Einstein's Theory of Relativity with a more sophisticated interpretation. Finally, Fitzgerald's idea stimulated the Dutch physicist, Lorentz.

Lorentz reasoned that the mass of a charged particle would increase as its length decreased. Assuming that the mass of a particle was due to the potential energy of its own charge, he argued that as the particle was crowded into a smaller space, its potential energy would increase and hence, its mass would increase. Since the potential energy is inversely proportional to the distance, he predicted the product of the mass and the length of the particle would remain constant and hence, the mass would increase according to the formulae $m = m_o/\sqrt{1 - v^2/c^2}$ where $m_o$ is the mass of the particle at rest and v is its velocity relative to the aether. This increase in mass was actually detected in 1900 by W. Kauffman. This result would also be incorporated in Einstein's Theory of Relativity five years later with a more satisfactory explanation.

Although the Lorentz–Fitzgerald interpretation of the Michelson–Morley experiment contained elements of Einstein's future theory of relativity, they represented a rear-guard action to preserve the notion of the aether. Einstein, unburdened by the traditions of the past, regarded the negative result of the Michelson–Morley experiment as evidence that the luminiferous aether of Maxwell simply did not exist. It is likely that he was aided in arriving at this conclusion in 1905 by the results of his work on photoelectric effect in the same year. In 1900, Planck had discovered that, apparently, the energy of light was quantized in discreet bundles of energy called photons. Einstein's work on the photoelectric effect in 1905 corroborated this view and showed that light in the form of photons sometimes behaves like a particle. We shall discuss this result in greater detail in Chapter 18.

The possible significance of Einstein's view that light can also behave like a particle is that the concept of an aether is no longer necessary for understanding the propagation of light through a vacuum. Light can propagate in the form of particles, and hence, the whole notion of an aether can be jettisoned. This is exactly what Einstein did. He went further. He also concluded that the light that traveled along the two different paths of the Michelson–Morley interferometer arrived at the same time because the velocity of light remained constant over the two paths and did not change as a result of the Earth's velocity. This deceptively simple interpretation of the Michelson–Morley result forms the foundation of Einstein's special theory of relativity: the velocity of light in free space is the same in all directions of space and is independent of the motion of the light source or the observer. This extremely revolutionary idea violates our intuitive notions of space, time and motion.

Let us first consider the addition of velocities. If I were to throw a ball from a moving car in the same direction as the motion of the car, then we all know from our own experience that the velocity of the ball is greater than the same ball thrown with the same force from a stationary car. It is common knowledge that the velocity of the ball and the car add. Is the same thing true of the velocity of light emitted from the headlights of our car? If we were to carry over our experience from the ball thrown from the stationary car and the moving car, we would surely think that the light from the headlights of the moving car would have a greater velocity than the light from the headlights of the stationary car. This conclusion, which we have reached intuitively, can also be supported logically by arguing in the following manner: If we must add the velocity of the car to the velocity of the ball, then we should add the velocity of the car to the velocity of light. The conclusion that the velocities of the car and light should add, nevertheless, contradicts the results of the Michelson–Morley experiment. Although this contradicts both our logic and intuition, it becomes apparent the one way to explain the negative result of the Michelson–Morley experiment is to assume that the velocity of light is independent of the velocity of the source of the light or of the observer.

The reason that it was so difficult for Fitzgerald, Lorentz, and their contemporaries to arrive at Einstein's simple, literal interpretation of the Michelson–Morley experiment was that it violated their intuition. For them, Einstein's conclusion wasn't natural. We experience the same discomfort in understanding and accepting Einstein's Theory of Relativity for it violates our intuition as well. His theory is based on the apparent paradox that the velocity of the car plus the velocity of light is equal to the velocity of light. But if Einstein's Theory of Relativity violates our intuitive notions of space, time and motion, why should we accept his interpretation rather than the Lorentz–Fitzgerald one? The reason is simply that the Lorentz–Fitzgerald interpretation represents a dead end. It is an ad hoc explanation of the Michelson–Morley experiment, which, aside from the increase of mass, did not make any other predictions, which could be tested experimentally. Einstein's interpretation led to a theory, which made a number of measurable predictions. All of the experimental tests that have so far been performed have all corroborated Einstein's Theory of Relativity.

Since the purpose of a scientific theory is to explain empirical observations rather than reinforce our intuitions, the choice of interpretation is obvious. We must learn to live with what seems to us an apparent paradox. If we consider for a moment, however, that our intuition concerning the addition of velocities developed only through the consideration of velocities, much less than the velocity of light, perhaps we may resolve this paradox. After all, the greatest velocity, any of us experienced was the speed of an airplane that is less than the speed of sound, which is only 300 m/sec or approximately one-millionth the velocity of light. Perhaps we may grant now that it is possible that the intuition we developed with super-slow velocities, like the speed of sound, do not apply when we are dealing with the velocity of light.

While this reasoning does not completely resolve the paradox in our minds, it helps us to understand how it is that the physics with velocities of the order of c can be so different from the low velocity physics that we intuitively identify with because of our limited experience. Bearing this in mind, perhaps we will not feel so uncomfortable with the ideas of relativity. Some discomfort is inevitable, however. We will notice that certain relativistic effects such as the Fitzgerald contraction encountered earlier disappear as the velocity involved becomes small compared with the velocity of light. For example, a mass of a particle traveling at the speed of sound increases only one part in $10^{12}$ and a length in the direction of motion shrinks by a similar amount.

# Chapter 14

# The Special Theory of Relativity

The Special Theory of Relativity is based upon the assumption that the velocity of light is independent of the motion of both the observer and the source of light. With this assumption, Einstein destroyed the notion of absolute motion. Before he formulated his Theory of Relativity, physicists believed that there existed a frame of reference absolutely at rest and that the motion of all objects in the universe was to be taken with respect to this absolutely stationary frame of reference. It was in this frame of reference that the aether sat motionless and, as a consequence, it was in this frame and only in this frame, that the velocity of light was exactly equal to c. If an observer was in a frame of reference and she wanted to determine the absolute motion of her frame of reference, all she had to do was to measure the velocity of light in her frame of reference. If the velocity of the light in her frame was exactly equal to c, then her frame was absolutely at rest. From the deviations of this velocity from c, she could determine the absolute motion of her frame of reference with respect to the stationary aether. This was the aim of the Michelson–Morley experiment, which as we know failed to detect any motion of the Earth with respect to the aether.

The negative result of the Michelson–Morley experiment led to Einstein's formulation of the Theory of Relativity in which the velocity of light is the same in all frames of reference independent of their motion. It is, therefore, impossible to determine the absolute motion of a frame of reference since the velocity of light will always be the same, namely c. All motion is relative. No frame of reference is preferred over any other. Einstein formulated this concept in terms of his principle of relativity, i.e. the laws of nature are identical in all uniformly moving frames of reference.

By a uniformly moving frame, we simply mean a frame of reference, which is not undergoing acceleration. In other words, the laws of nature are identical in a frame that is stationary with respect to the Earth and in the frame that is in the interior of a train moving at a constant velocity with respect to the Earth. (We are ignoring, for the purposes of this discussion, the motion of the Earth in its orbit about the Sun or its rotation). If the train was accelerating or decelerating, the laws of physics in the frame of the train would not be the same as those in the frame at rest with respect to the Earth. The reason for this is a mass with no force acting upon it would experience a fictitious force in the accelerating train but not in the frame at rest with respect to the Earth. The fictitious force arises because the mass will continue to move at a constant velocity due to its inertia independent of the motion of the train. If the train accelerates then the constant motion of the mass will no longer be constant with respect to the accelerating train. With respect to the train, the mass will appear to be accelerated by a fictitious force. This is a familiar experience to all who have been thrown forward in a moving train or bus, which suddenly decelerates. When describing the laws of physics it is, therefore, wisest to remain in a non-accelerating frame of reference so as to avoid the presence of friction forces. In our discussion of special relativity, therefore, we will automatically limit ourselves to frames of references for which there are no fictitious forces. Therefore, unless otherwise mentioned, the reader may assume that the frame of reference under discussion is one undergoing uniform motion. Einstein's principle of relativity states that all such frames are equivalent and the laws of physics described in each of these frames are identical.

The idea that all motion is relative was as devastating to Einstein's contemporaries as Copernicus' notion that the Earth was no longer the center of the universe. It took people over one hundred years to accept the idea that the Earth actually moved. Once this idea was accepted, thinkers became used to the idea that the Earth was moving with respect to some frame of reference absolutely at rest. So, although the Earth was not at rest, at least there was some place in the universe at rest, a place where one could anchor one's thoughts. The concept of absolute motion and absolute space, which evolved essentially from the physics of Copernicus, Galileo, and Newton was elevated to the heights of an *a priori* truth by the philosopher, Immanuel Kant. This is an indicator of how engrained the concept of absolute motion became in the minds of Western thinkers. Kant could not conceive of the possibility that space could be structured in any other way. Einstein's principle of relativity, based on experimental fact, completely destroyed the validity of this so-called *a priori* or absolute truth. Instead of there being one frame of reference at rest, Einstein showed that any frame of reference not accelerating may be considered at rest and the motion of all the other objects in the Universe may be taken with respect to it.

We have all experienced the relativity of motion while riding in a train or a subway car. Remember the sensation when your car was at rest in the station and you looked out the window to see another car besides yours moving out of the station. For a moment it feels as though your car is in motion. In fact, unless you can observe the ground or some other fixed structure in the station it is impossible by just looking at the other car to determine whether your car or the other car is in motion. This example illustrates the relativity of motion. Before Einstein, physicists believed they could always tell which car was really in motion. Einstein has shown this is pure folly. You can determine which car is moving with respect to the Earth but there is no way of determining whether the Earth and the stationary car are really at rest and the moving car is in absolute motion or vice-versa.

In the Einsteinian universe, the whole notion of space changes; it is no longer defined in the absolute sense in which it is in the Newtonian universe. For Einstein, space does not exist per se. Space is a relationship between physical objects. If there were no objects in the universe, there would be no space in Einstein's universe, whereas in Newton's universe, there still would be space. Space has an *a priori* existence for Newton. It is the container, which holds the universe. If one considers a finite universe from the Newtonian point of view one can discuss the space outside the universe whereas in the Einsteinian worldview, one cannot. Space for Einstein is a relationship, whereas space for Newton is something, which has existence and reality.

Let us illustrate their two contrasting viewpoints by considering the space between my two outstretched hands. What happens to that space when I put my hands down at my side? From the Newtonian point of view, the space in front of me is still there, just sitting there ready to be filled, if necessary. From the Einsteinian point of view, once my hands have disappeared, we can no longer talk about the space between them or the space that was between them. The reasoning is similar to the one employed in the riddle: Where does my lap go when I stand up? My lap is the relationship of my legs when I am seated and no longer exists when I stand. In a similar fashion, according to Einstein's way of thinking, space is also a relationship of physical objects, which disappears when those objects disappear.

The resistance to the adoption of Einstein's ideas was due to the radical change required in the way one regarded space. This, plus the non-addition of the velocity of light, made his theory very unappealing to his contemporaries. The Einsteinian worldview was also unpopular because of the way in which time was regarded. In the Newtonian worldview, time had an absolute status, like space. In the minds of scientists and philosophers, there was a clock somewhere in the universe, which went on ticking through eternity in terms of which all events took place. In fact, one of the popular images of the universe was that it was a giant clock, which kept an eternal time. The absolute nature of time, like space, was also given the status of an *a priori* truth in Kant's philosophical system.

In the Einsteinian worldview, on the other hand, time is merely the relationship between events. It does not exist per se. If there were no events in the universe, then there would be no time. We have all, no doubt, experienced this concept psychologically. When nothing is happening to us, we lose our sense of time. Those who have spent long periods of time underground with nothing to do lose their sense of time altogether. We are most aware of time when we have a great number of things we must do. Let us consider a universe with a finite lifetime so that there are two events corresponding to the birth and death of the universe. Within the framework of Einstein's worldview, one cannot conceive of or discuss the time before the creation of the universe or the time after its destruction. These times simply do not exist as they do in the Newtonian worldview. As with space, time for Einstein is a relationship, whereas, for Newton, time is something, which exists and is real. Time, in the Newtonian sense, is that which a clock tells. Just as a clock, if properly wound, would continue to run, so time for the Newtonian continues to pass.

In order to develop a feeling for time in Einstein's world, we should examine how the concept of time developed before the existence of the clock. In essence, our sense of time developed from our ability to order or sequence events. If I were to whistle, stamp my feet and then clap my hands before a group of students, there would be a general agreement as to the order in which I performed each of the events. There might be disagreements as to the duration between events, but there would be universal accord concerning the sequencing of the events.

This ability for us to sequence events provides the foundation for our sense of time. As man began to associate the events in his life with the motion of astronomical bodies such as the Sun and the Moon, his concept of time became more sophisticated. He began to describe the intervals between events in terms of the number of times the Sun had risen between the two events or in terms of the number of moons. Later, he divided the interval between successive sunrises into smaller units called hours. Each hour corresponded to a different position of the Sun in the sky. He began describing the intervals between events in terms of the change of the Sun's position in the sky.

Finally, instead of always relating events to the motion of the Sun or Moon, man created an instrument called the clock, which, like the Sun in the sky, executed a regular type of motion. A correspondence was then made between events in the real life and events on the clock. So, for example, when we say that we arose at 8 o'clock in the morning, what we are in essence saying is that the event of our arising and the event of the hands of the clock indicating 8 o'clock coincided. If we think of a clock as an instrument, which performs a regular and repetitive motion to which we may refer other events, we will be more in the spirit of the Einstein view of time. If we consider the clock as an instrument, which measures the passage of time, we shall retain the Newtonian concept of time.

The Newtonian concept of absolute motion is destroyed in the Theory of Relativity because the velocity of light is a constant in all frames of reference. As a consequence, all motion is relative and space is no longer absolute as it was in Newtonian physics. It is also the constancy of the speed of light, which destroys another concept central to Newtonian physics, namely, the absoluteness of time. According to Newtonian thinking, there is only one absolute time, which is valid for all frames of reference. In the Theory of Relativity, on the other hand, time is also relative to the particular frame of reference under consideration. In other words, the time in two different frames of reference is not the same.

In order to illustrate how this follows from the constancy of light, we will consider a hypothetical situation originally suggested by Einstein to explain his ideas. In essence, we will conduct a thought experiment, that is, create an experimental situation in thought. Let us consider two observers, one sitting on a train moving at a constant velocity with respect to our second observer sitting at rest along the track on which the train is traveling as shown in Fig. 14.1. Two flashes of lightning strike the front and the rear of the train when the train is in the position represented by the dashed lines. The train has a length L and velocity v.

The observer, Dr. A, sitting precisely in the middle of the train sees the two flashes of lightning hit the train at exactly the same time. He is sitting exactly in the middle of the train and, therefore, the times for the transit of the signals of the lightning hitting the two ends of the train to his eyes are the same, namely L/2c. He concludes that the two flashes of lightning struck the train at exactly the same time.

The train travels a distance of vL/2c in the time it takes the light from the lightning flashes hitting the train to travel into Dr. A's eyes. The train at this moment is represented by the solid lines. An observer, Dr. B, is stationed at rest beside the railroad track so that he is exactly opposite the moving observer, Dr. A, precisely at the moment when the light from the lightning striking the ends of the train enters Dr. A's eyes. Will the stationary observer, Dr. B, also report that the two bolts of lightning struck the train at the same time? In order to determine this, let us consider how Dr. B will perceive the two events under consideration. When the lightning actually struck the train, the train was farther down the tracks by the amount vL/2c, as illustrated in Fig. 14.1. This is the distance between the position of the dashed train when the lightning strikes and the position of the solid train when Dr. A sees the two bolts of lightning at the same time. Note that at the moment the lightning struck the train, the stationary observer, Dr. B, was actually closer to the front of the train than to the rear of the train. Because of the fact that the velocity of light is a constant, the light from the two bolts of lightning striking the front and the rear of the train travels towards our stationary observer, Dr. B, at the same speed, c, even though the front of the train is moving away from him and the rear of the train is moving towards him. Because of the constancy of the velocity of light and the fact that the stationary observer is closer to the front of the train when the lightning strikes, he will see the lightning strike the front of the train before it hits the rear of the train. He will conclude that the lightning struck the front of the train before it hit the rear of the train. The stationary observer, Dr. B, has a completely different sense of time from that of the moving observer, Dr. A. The two events, which the moving observer, Dr. A reported to be simultaneous, occur at different times in the frame of reference of the stationary observer, Dr. B.

The discrepancy in the timing of the two events by our two observers is due to the fact that the velocity of light is a constant in both frames of reference. If the velocity of light added with the velocity of the train, as our intuition would have us believe or as Newtonian physics would instruct us, then the velocity of the signals arriving from the front and the rear of the train to the stationary observer, Dr. B would not have been the same. The velocity of the light from the lightning striking the rear of the train would have traveled towards the stationary observer with the velocity c + v while the signal from the front of the train would have traveled with the velocity c – v as Fig. 14.1 indicates. The distance from the rear of the train to the stationary observer when the lightning strikes is L(1 + v/c)/2 whereas the distance to the front of the train is L(1 – v/c)/2. This means that if the velocities had added in the Newtonian sense, they would have arrived at exactly the same time. We must, therefore, attribute the different sense of time possessed by our two observers to be due the constancy of the velocity of light.

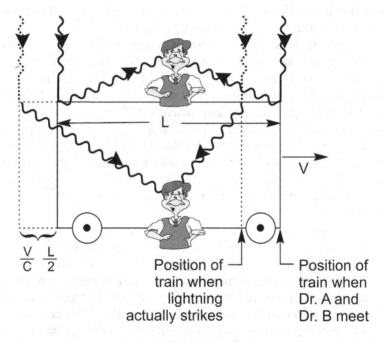

Position of ⌐ ⌐ Position of
train when train when
lightning Dr. A and
actually strikes Dr. B meet

Fig. 14.1

The result of the constancy of the velocity of light will also affect the spatial perceptions of the two observers moving with a constant velocity with respect to each other. The difference in the spatial perception of our two observers manifests itself in terms of the Lorentz-Fitzgerald contraction. Let us consider a stick, which, at rest, has the length $L_o$, moving with respect to a stationary observer with the velocity v parallel to its length. The stationary observer will discover that, in his frame of reference, the stick has suffered a contraction and only has the length $L_o \sqrt{1 - v^2/c^2}$. An observer moving with the velocity, v, of the stick will discover, however, that in his frame of reference, the stick, at rest, still has the length $L_o$. In other words, as a result of its motion, an object appears to a stationary observer to have contracted by the Fitzgerald factor, $\sqrt{1 - v^2/c^2}$, which always has a value less than one but greater than zero. The contraction of the object only takes place in the direction parallel to its motion. The length of the object perpendicular to its motion will appear the same to the stationary observer. A moving square will appear, to the stationary observer, to shrink into a rectangle whose shorter side is parallel to its motion and a moving circle will become an ellipse with its major axis perpendicular to its motion.

When the meter stick is at rest its length will be observed to be one meter. When the velocity of the stick is 0.5c then its length, to the stationary observer, is observed to be 0.88 meters, as its velocity increases to 0.88c, its length now appears to be only one-half of its original length. When its velocity reaches 0.999c, then its length will appear to be only 0.045 meters. As the velocity of the stick approaches closer and closer to the velocity of light, its length appears shorter and shorter to the stationary observer. An object traveling at the velocity of light would literally disappear from view. This is impossible, however, since no particles with mass can travel faster than or at the velocity of light, as we shall soon see.

The interpretation of the Lorentz–Fitzgerald contraction within the framework of the Theory of Relativity is quite different from the one originally made by Lorentz and Fitzgerald. According to these early workers, the moving meter stick actually undergoes a physical contraction due to pressure of the aether wind arising from the absolute motion of the meter stick with respect to the aether. The contraction is due to the absolute motion of the meter stick. From their point of view, if the observer were to move and the meter stick remained stationary, the observed would not observe a contraction. In the Theory of Relativity, on the other hand, the contraction is observed whether the stick moves with respect to the observer or the observer moves with respect to the stick. Furthermore, the stick does not actually physically contract. An observer moving with the stick does not observe a change in its length. The stick appears shorter to the stationary observer because his concept of space is different than that of the moving observer. They perceive the same physical phenomenon differently.

Not only are the spatial perceptions of the stationary and moving observers different but their temporal perceptions differ also. We saw evidence of this earlier when we considered the example of the lightning striking the train. The two observers had a different notion of the simultaneity of the lightning flashes. Perhaps even more surprising is the fact that a stationary observer will observe that time in a moving frame of reference actually slows down. A moving clock runs slower than a stationary clock.

For example, if a clock is moving at the velocity of 0.866c with respect to a stationary observer, the time required for the passage of one hour on the moving clock will take two hours on a clock at rest in the stationary frame. The moving clock appears to be losing time or slowing down. In general, a time interval that takes $t_0$ seconds in the moving frame of reference will take $t = t_0 / \sqrt{1 - v^2/c^2}$ seconds in the stationary frame. The time elapsed in the stationary frame for the passage of one hour of a moving clock is always greater than one hour. Notice that, as the velocity of the moving clock approaches c, the time for one hour to elapse in the moving frame, requires an infinite amount of time in the stationary frame or, in other words, time comes to a standstill in the moving frame from the point of view of the stationary observer. Remember that it would be impossible for the moving clock to achieve a velocity c with respect to a stationary observer because of its finite mass.

The slowing down of time in a moving frame of reference must seem like science fiction to the average reader who has had no experience with objects moving at velocities any where near the velocity of light. To an elementary particle physicist who deals all the time with sub-atomic particles, which travel at velocities near the velocity of light, the slowing down of time in a moving frame of reference is a very real thing. It, in fact, helps him in his study of certain short-lived elementary particles called mesons. Mesons are created with a high-energy accelerator as the result of the collision of protons with other protons or nuclei. One of the mesons created in such a collision is the short-lived $\pi$ meson, which lives, on the average, about $10^{-8}$ sec before decaying into other elementary particles. The average lifetime of $10^{-8}$ sec quoted for the $\pi$ meson is the value of the meson's lifetime when it is observed at rest. As the velocity of the meson, with respect to the observer in the laboratory increases, so does the lifetime of the meson.

The reason for this is that time for the moving meson, like a moving clock, slows down and hence, the lifetime of the meson at rest, $t = 10^{-8}$ sec, takes longer to pass in the laboratory frame. In fact, the measured lifetime of the meson, as a function of its velocity, is just $t = 10^{-8}/\sqrt{1 - v^2/c^2}$ sec, exactly the value one would predict on the basis of the Theory of Relativity. The extra time that the meson lives in the laboratory frame, as a result of its velocity, has a practical aspect. It gives the elementary particle physicists enough time to study the meson before it decays. If it was not for the time dilation of the meson's lifetime, the meson would decay after $10^{-8}$ sec and there would not always be enough time to study it in the laboratory frame.

The slowing down of time in a moving frame of reference is not just a theoretical idea but something, which occurs daily in the laboratory of the elementary particle physicists. Although the time in the moving frame appears to slow down for the stationary observer, an observer in the moving frame does not experience time slowing down. As far as the moving observers are concerned, their clocks run at the same rate. An hour is an hour and the lifetime of a $\pi$ meson is only $10^{-8}$ sec. This is to be expected since, from their point of view, they are at rest and hence, expect time to flow naturally. In fact, from the point of view of the moving observers the stationary frame of reference is in motion with respect to them, and therefore, they observe time slowing down in this "so-called" stationary frame, which they interpret as moving. Both observers in the two frames see the clocks in the other frame as slowing down.

As paradoxical as this may seem, both the stationary and the moving observer will observe each other's clock slowing down. Each believes that the time is dilated in the others' frame. How is this possible? How can time slow down for both? In addition to this mutual time dilation, each observer will also observe the others' meter stick to be shorter. Is this possible? Which meter stick is really shorter? Which clock is really telling the correct time? We cannot ask questions like this, because in asking such a question, we are assuming the existence of the absolute space and absolute time, which characterized Newtonian physics. Within the framework of Einsteinian physics, we cannot discuss such things as the real length of the stick or the real time of the clock. The observations of the observers in the moving frame are as real and as valid to them as the observations of the observers in the stationary frame are to them. We simply must accept the notion that two observers moving with respect to each other do not empirically observe the same things. The unstable meson only lives $10^{-8}$ sec in the frame in which it is at rest and it really lives $10^{-8}/\sqrt{1-v^2/c^2}$ sec in the stationary frame in which its velocity is v. The quantities that observers measure are only relative to their particular frame of reference. One cannot make an absolute comparison. Both observers will really observe the other's clock as slowing down.

The phenomenon of relativistic time dilation might, some day, make interstellar space travel possible. The closest star, Alpha Centauri, is four light years away. A light year is a unit of length equal to the distance that a beam of light travels in one year or about nine trillion kilometers. This means that, since the velocity of light is the fastest speed attainable no matter how proficient our technology became, it would take a minimum of eight years to send an astronaut to Alpha Centauri and back. Since there are only a few stars in this corner of our Milky Way galaxy that are within fifty light years, the minimum time to send a space craft to most stars would be 100 years or more, which is greater than the lifetime of even our healthiest astronauts. Even if we were to develop a space craft which could travel at velocities near the speed of light, it does not seem that it would be possible to send a man to a distant star and have him live long enough to return alive.

Because of the relativistic time dilation, however, this is actually possible. The reason for this is that, while the astronaut is traveling to a distant star at a velocity close to c, his clock would slow down with respect to an Earth clock. So, for example, if he was traveling at 0.999c to a star 99.9 light years away, he would require 100 Earth years to make the journey to the star. In his own frame of reference, however, he would age only 4.47 years. (These two times are related by the formulae t = $t_o/\sqrt{1-v^2/c^2}$, where t = 100 years, $t_o$ = 4.47 years and v = 0.999c).

One might wonder how it is possible for the astronaut to travel 99.9 light years in only 4.47 years. The reason is, that in his frame of reference, the distance to the star is contracted by exactly the same factor that his time is dilated with respect to the Earth and therefore he sees that the star is only 4.46 light years away and is traveling towards him at 0.999c . Therefore, it takes only 4.47 years for the journey to the distant star.

The astronaut will only require 8.94 years, according to his clocks, in order to make the round trip to the star and back again. However, when he returns to Earth, he will discover that, during the approximately nine years he was away according to his clock, 100 years had elapsed on

Earth. The astronaut will have aged only nine years but all of his friends will have aged 100 years and hence, are all dead. The astronaut will have experienced not only a round trip to a star 100 light years away, but he will also have experienced traveling 91 years into the future. The exploits of our hypothetical space traveler have given rise to a famous paradox known as the twin paradox. Let us assume that our space traveler has a twin brother who remains on Earth during his flight. The twin on Earth will expect his brother in space to age at a slower rate as a result of his motion, as we explained above. The formulators of the twin paradox maintain, however, that the space traveler will also expect his brother on Earth to age at a slower rate. According to their argument, during his flight the astronaut, in his frame of reference, will see his brother on Earth moving away from him and then towards him. As a consequence, he will observe his brother's clock slowing down and hence, he, too, will expect his brother on Earth to age at a slower rate than himself.

One is faced with a paradox. Which brother will actually be the younger when the two brothers are reunited back on Earth? The brother who actually traveled to the star and came back to Earth will be the younger. In resolving the paradox, we must take into consideration the effect of the traveler's motion on his observations. The paradox is only an apparent one because it is based on an assumption, which is not valid. When we considered the astronaut's observations of the Earth-bound clocks, we assumed that his observations were made in a frame of reference in uniform motion. This assumption is not valid because, in order for the astronaut to return to Earth, he must change directions in the middle of his trip. This naturally involves acceleration or non-uniform motion.

The astronaut was no longer in a non-accelerating frame of reference during the time that he is turning around and as a result, his observations of time on the Earth are affected. It is true that, while he is traveling at a constant velocity, he observes the Earth clocks slowing down with respect to his. During the time that he is turning around, however, he observes the Earth clocks as speeding up. Although the time, in his frame to reverse directions might not take very long, it is during this period that he observes the passage on Earth of 91 years that will separate the earthbound clocks from his clock when he returns to Earth.

The twin paradox has also been resolved experimentally. We mentioned earlier that mesons, which are moving at a constant velocity, take longer to decay in the laboratory frame. The increase of the laboratory lifetime, because of uniform motion, is also observed for the short-lived π meson, an unstable elementary particle that decays at rest in $10^{-8}$ sec. When the π meson is accelerated in a circular accelerator, it is found that its lifetime in the lab frame increases in the same manner as its lifetime in the lab increase when it is in uniform motion. The π meson, which makes a round trip around the accelerator, comes back younger than a π meson, which remains at rest in the laboratory. In other words, the π meson twin, which actually makes the journey, comes back younger.

Now that we have experimental proof that a space traveler will come back younger than his twin, we know how to build a time machine. H.G. Wells, in his famous short novel, The Time Machine, describes a device that allows its passenger to travel into the future. When this story was first written, the idea of a time machine was pure fantasy. This idea is no longer a fantasy. The machine, which accelerated the π meson, is a time machine, which projects the π meson into the future. If the π meson had not been accelerated, it would have decayed within $10^{-8}$ sec, like all the other π mesons, which were created at the same time as it was. Instead, as a result of its motion, it found itself still in existence long after the time it would have naturally ceased existing if it were at rest. It was existing in the future, i.e. the time after it was suppose to have decayed.

If it were possible to build a spacecraft that allowed a man to travel at velocities near the velocity of light, then one could travel into the future by making a round trip journey. The closer to c at which one travels; the farther into the future one would be projected. For example, if one wished to devote one year of his life to traveling into the future and one's space ship would travel at the velocity v, then he would find himself projected $1/\sqrt{1-v^2/c^2}$ years into the future. If he traveled at v = 0.99995c he would be projected 100 years into the future whereas, and if he traveled at v = 0.999999995c, he would be projected 10,000 years into the future. If one was unhappy about the point in time one arrived at, one could always get back into the space ship and travel a bit further.

The only problem with our time machine, however, is that once one traveled into the future, it would not be possible to travel back to the time one started from. H.G. Wells' time machine did not suffer from this problem. His device was purely a product of his imagination. Our device, while not practical at the moment, does not violate any of the laws of physics, however. Science fiction writers, since the advent of the Theory of Relativity, have exploited the relativistic time machine, described above, to project their characters into the future. In order to project backwards in time, these writers usually have their character travel faster than the speed of light, a violation of the Special Theory of Relativity.

But how does traveling faster than the speed of light project one backwards in time? It really cannot. If one could travel faster than the speed of light, one could overtake light signals that were emitted from Earth a long time ago in the past. By intercepting these signals, one could observe events that occurred hundreds of years ago such as the building of the pyramids or the extinction of the dinosaurs. While this is not equivalent to actually living in the past, it represents traveling into the past in a certain sense. One can see how a science fiction writer might be able to imagine his character traveling backwards in time. Unfortunately, it is not possible for us to travel faster than c, so there is no hope for seeing back into the past. We can, however, anticipate traveling into the future but this will require tremendous advances in space travel technology. At the moment, we can only send $\pi$ mesons into the future.

Our claim that man cannot travel faster than the speed of light is based on experimental facts. Accelerators have been built, which accelerate electrons and protons through electric potentials. The velocity of the charged particles increases as a result of their absorbing electrical energy. It is found that, as the velocity of the charged particle approaches c, a large amount of energy is required to increase the particles' velocity by even a small amount. One finds, in fact, that no matter how much energy is transferred to the particle, its velocity never reaches c. Apparently, the velocity of light is the ultimate velocity of the universe. Since all matter is composed of particles, such as electrons and protons, these experimental results show that the velocity of light may never be exceeded.

In order to understand why particles cannot go faster than c, it is necessary to examine the behaviour of the mass of a particle as a function of its velocity. In fact, the mass of the particle becomes infinite as the velocity of the particle approaches c. Since the mass of a particle depends on its velocity, it is useful to define its rest mass, which is the value of its mass when it is at rest, and which we denote by $m_0$. The mass of a moving particle, with respect to a given frame of reference, is related to its rest mass, $m_0$ and its velocity v, with respect to that frame by the formula, $m = m_0/\sqrt{1 - v^2/c^2}$. As the velocity of a particle approaches c its mass becomes infinite to a stationary observer.

It is the increase of mass with velocity, which explains why the ultimate or highest velocity of a particle is c. The mass of a particle is a measure of its resistance to acceleration by a force. As one accelerates a particle to velocities approaching c, its mass increases significantly making it more difficult to accelerate. Because the mass becomes infinite, exactly in the limit that the velocity becomes c, it is impossible to generate an infinite force. This explains why the ultimate velocity of matter is c. No massive particle actually obtains the velocity c since this would require an infinite amount of energy. Particles whose rest mass is zero, however, such as the photon, travel at the velocity c. In fact, photons can never have a velocity less than c.

The expression for the energy in the Theory of Relativity differs quite markedly from its expression in Newtonian physics. Einstein found, however, that there is a simple relationship between the energy and the mass of a particle, namely, the energy of a particle is equal to its mass times the velocity of light squared, $E = mc^2$. If we express the relativistic mass in terms of the particle's rest mass, $m_0$, and velocity, v, then the familiar expression for energy, $E = mc^2$ becomes $E = m_0c^2/\sqrt{1 - v^2/c^2}$.

Einstein's expression for the energy for velocities much less than the speed of light, reduces to $E = E = m_0c^2 + 1/2m_0v^2$, which differs from the non-relativistic expression for the kinetic energy, $1/2m_0v^2$, by the addition of the extra term, $m_0c^2$. This extra term is the rest mass energy of the particle and is equal to the energy of the particle at rest. According to the interpretation of this term given by Einstein, the particle has energy simply by virtue of its rest mass.

This means that mass and energy are equivalent. If this is true, then one should be able to convert mass into energy and energy into mass just as one converts energy into heat and heat into energy. This was one of the major predictions to grow out of Einstein's Theory of Relativity. It was on the basis of this prediction that work on the construction of the atom bomb began in the U.S.A. during World War II. It is sad that the prediction of a man who struggled for world peace, as Einstein did, should be verified in a device of war and destruction such as the atom bomb. The atom bomb was indeed the first device to convert mass into energy on a large scale. Nuclear fission had only been observed in laboratory situations before this in which only tiny amounts of energy were generated.

The energy released in nuclear fission results from the destruction of matter in the splitting of Uranium 235 into Barium and Krypton and 3 neutrons. The total mass of all of the products of the fission reaction is less than the masses of the original uranium nucleus and the neutron, which triggered the reaction. Another example of a nuclear reaction in which energy is released is thermo-nuclear fusion, the process responsible for the energy of the Sun. The mass of the final product of fusion, the helium nucleus, is less than the sum of the masses of the hydrogen nuclei, which combined or fused together to form the helium nucleus. In both nuclear fission and fusion the mass that is destroyed in these reactions is converted into pure energy.

Another process whereby energy is converted into matter and back into energy is the process of pair creation and annihilation. In this process, energy, in the form of light, is converted into the masses of an electron and an anti-electron. An anti-electron or positron is a particle with the same mass and spin of an electron, but which has the opposite charge. An anti-electron, like an electron, is stable by itself. However, when an electron and anti-electron make contact, they annihilate each other and change back into light energy. The creation and annihilation of electron-positron pairs is a process whereby light energy is converted into matter. When an anti-electron and an electron collide they annihilate each other and their mass is converted back into pure energy in the form of photons. We will study this process in greater detail when we come to our study of elementary particles at which time we shall encounter other examples, which demonstrate the equivalence of mass and energy.

Before completing our study of the Special Theory of Relativity, we turn to one of its first applications in which Einstein demonstrated the equivalence of the electric and magnetic forces. Maxwell had shown, through his equations, that the electric and magnetic fields were intimately connected to each other. Einstein was able to demonstrate the actual equivalence of these two forces, however, by using his principle of relativity, which states that the laws of physics for two observers moving at a constant velocity with respect to each other are the same.

Einstein asked us to consider two charged particles moving along two parallel lines at constant speed with respect to a stationary observer. The stationary observer will observe two forces between the charged particle. One force, the electric force, arises solely by virtue of their charge. The other force, the magnetic force, arises by virtue of both their charge and their velocity. From the point of view of an observer moving with the same velocity as the charged particles, there will only be an electric force between the charges. The observer will not observe a magnetic force because, with respect to the observer, the charges are both at rest. The interaction of the charged particles is identical for the two observers because of the principle of relativity. Hence, the electric and magnetic force observed by the stationary observer is equivalent to the electric force observed by the moving observer or, in other words, the electric and magnetic forces are the same. The magnetic force is just an electric force in motion.

# Chapter 15

# The General Theory of Relativity

Einstein's Special Theory of Relativity deals with the effects of uniform motion on the relative space and time perceptions of moving and stationary observers. Discussion within the Special Theory of Relativity was always limited to non-accelerated motion. As soon as accelerated motion was encountered, as in the twin paradox, we found ourselves beyond the scope of the Special Theory of Relativity.

It is in Einstein's General Theory of Relativity, published in 1916, that accelerated motion is taken into account. The General Theory of Relativity generalizes the results of the Special Theory and deals with the effect of accelerated motion on space and time perception. The General Theory also deals with the effects of a gravitational field on space and time perception by showing the equivalence of accelerated motion and the existence of a gravitational field. This is accomplished by exploiting the fact that the inertial and gravitational masses of all particles are equal. Finally, the General Theory of Relativity provides a theory of gravity, which replaces Newton's theory of gravity.

We discovered that, within the framework of the Special Theory of Relativity, it is impossible to determine the absolute motion of a frame of reference through space as long as that frame is not undergoing acceleration. Once a frame of reference undergoes acceleration, however, it is possible to detect the acceleration because of the fictitious forces, which arise as a consequence of the acceleration. These forces are fictitious because, in actuality, the acceleration of a body, with respect to the frame of reference that these fictitious forces seem to cause, is only apparent. In actuality, the body continues to move with constant velocity because of inertia but, relative to the accelerating frame, it appears to be accelerating. The fictitious force causes real acceleration with respect to the accelerating frame, however, and hence, is perceived as a gravitational force. Think of how you have been thrown forward when the bus driver suddenly slammed on the brakes.

Einstein exploited this fact and claimed that the effects of the accelerated motion are equivalent to those of a gravitational force. In order to make this assertion, he noted that because the gravitational and inertial masses are the same, objects will accelerate at exactly the same rate in both a gravitational field and in an accelerating frame of the reference. He therefore concluded that accelerated motion and the action of a gravitational field are equivalent.

In order to illustrate this, let us consider the two elevator cars in Fig. 15.1. One elevator car is sitting at rest on the surface of the Earth. The other elevator car is situated in outer space completely removed from the effects of any gravitational field. The elevator car, however, is being accelerated upwards with a constant acceleration, g, equal to the acceleration of the particle falling near the surface of the Earth. We claim that an observer in one of these cars would not be able to determine in which car she was situated because her experiences in both cars would be identical. She would experience a gravitational pull towards the floor of the elevator in both cars.

Fig. 15.1

If the observer dropped two objects, let us say, an iron ball and a wooden ball, in the elevator car on Earth, these two balls would fall at exactly the same rate because of the equality of their gravitational and inertial masses. If our observer were to drop the balls in the accelerating elevator car in outer space, she would experience the two balls accelerating towards the floor at exactly the same rate as they did in the elevator car on Earth. The reason for this is as follows: as the observer, releases the two balls from her hands, the balls will no longer be accelerated upwards. They will remain at rest in a frame of reference, which was moving at the same speed as the elevator at precisely the time the two balls were released. The elevator continues to be accelerated, however, and hence, the floor of the elevator will accelerate up at the rate, g, towards the balls, until the floor hits the balls. To an observer in the elevator, however, it appears as though the balls had fallen to the floor with the acceleration, g. In other words to an observer in the accelerating elevator, the masses behave exactly as they would in the elevator at rest on the surface of the Earth. The acceleration of his elevator car, therefore, exactly duplicates a gravitational field.

If the direction in which the elevator car was being accelerated changed direction so that the elevator was being pulled down with an acceleration, g, then there would be an effective gravitational field pulling things towards the ceiling of the elevator instead of towards the floor. If such an elevator were placed near the surface of the Earth, then the effects of the gravitational field produced by its acceleration down would exactly cancel the gravitational field of the Earth. This is exactly what happens in an aircraft, which is experiencing free fall. The effective gravitational field created as a result of its being accelerated down by the Earth, exactly cancels the gravitational field of the Earth and hence, observers in the free falling craft experience no gravity. Objects will float around inside the aircraft just as they do when an astronaut travels to the Moon and is no longer within the influence of the Earth's gravitational field.

Following Einstein, we have established that the laws of physics describing the motion of massive particles are the same in the elevator car sitting in the Earth's gravitational field at rest and in the elevator car being constantly accelerated upwards with an acceleration, g. Einstein concluded that, on the basis of this demonstration, all the laws of physics would be identical in these two frames of reference. He expressed his hypothesis in terms of the equivalence principle, which states the following: the phenomena in an accelerating frame of reference are identical with those in a gravitational field.

The equivalence principle forms the heart of the General Theory of Relativity. Einstein exploited this principle to successfully predict that, a gravitational field would bend a beam of light, i.e. a beam of photons. This is somewhat surprising in view of the fact that the rest mass of a photon is zero. Although the rest mass is zero, the photon still has energy, and since Einstein showed that mass and energy are equivalent, perhaps light can also be affected by gravity. To demonstrate this, we turn to the propagation of light in an accelerating elevator car.

Let us consider a beam of light propagating perpendicular to the direction of the acceleration and entering the elevator from one wall and exiting the elevator on the opposite wall. The beam of light will appear bent to an observer in the elevator, as shown in Fig. 15.2. The reason for this is that, by the time the light beam has propagated from one wall to the other, the elevator has moved upwards because of its acceleration. The beam of light, however, is unaffected by the elevator's acceleration and hence, continues to propagate along the same straight line it was moving along before it entered the elevator. Relative to the accelerating elevator, however it appears to exit at a point below the one it entered. The beam of light will appear, to an observer in the accelerating elevator, to have been bent by the same gravitational field that causes the objects she drops to also fall to the floor. If, instead of a beam of photons, we had sent a beam of massive particles through the elevator, they would behave exactly like the beam of light. An observer in the accelerating frame will conclude that the paths of both the massive particles and the

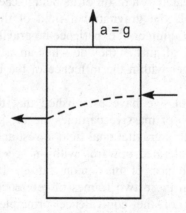

Fig. 15.2

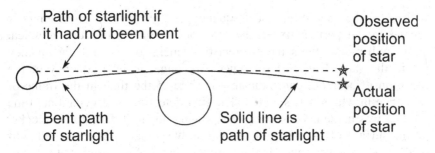

Path of starlight if it had not been bent

Bent path of starlight

Solid line is path of starlight

Observed position of star

Actual position of star

Fig. 15.3

light were bent because both the massive particles and the light were attracted by the gravity. At this point, Einstein exploits the equivalence principle. He argues that, since the laws of physics are the same in the accelerating frame and the stationary frame sitting in a gravitational field, then, one can expect the beam of light to be bent by a bona fide gravitational field such as that generated by the Sun.

On the basis of this argument, he predicted that starlight passing close to the Sun would be bent by its gravitational field and hence, during a solar eclipse, the position of a star close to the Sun in the sky would be displaced from its normal position. In Fig. 15.3, we illustrate how the bending of the starlight by the Sun makes it appear that the position of the star has changed position. The first opportunity to measure the effect of the gravitational pull of the Sun on starlight came during the total solar eclipse of 1919. The expedition of British scientists who traveled to Africa to observe the eclipse, found that the position of stars near the Sun had indeed changed, as Einstein had predicted. This measurement verified the validity of the equivalence principle.

Einstein further exploited the equivalence principle to determine the effects of a gravitational field on the space and time perceptions of an observer. Let us consider from the point of view of a stationary observer a clock in an accelerating elevator car. By virtue of the fact that the clock develops a velocity with respect to the stationary observer, as a result of the elevator's acceleration, we expect the clock to slow down. Hence, the stationary observer will observe a time dilation in the accelerating frame of reference. By virtue of the principle of equivalence, we would also expect that a clock sitting in a gravitational field would also slow down.

This gravitational time dilation has been experimentally verified in two separate experiments, both of which involve 'atomic clocks'. An atom is like a clock. The electron orbits the nucleus of the atom with a fixed periodicity. The frequency of light that an atom emits is related to the periodicity of the electron's orbits. Since two identical atoms will radiate the same discreet frequencies or spectral lines if they are in the same frame of reference, a change in the frequency of the radiation of an atom can indicate a change of the time in the frame of reference in which it is situated. The first detection of gravitational time dilation was made by observing the slowing down of the 'atomic clocks' on the surface of extremely dense white-dwarf stars. The expected shift of the spectral lines towards the red was observed. A similar effect was observed for the spectral lines emitted from the surface of the Sun. Pound in a laboratory setting at Harvard University made a more precise observation of the gravitational time dilation. Pound compared the spectral lines of two atomic clocks separated by a mere 20 meters in the Earth's gravitational field. He found that, the 'atomic clock' sitting closer to the Earth and hence, more strongly influenced by its gravitational field, slowed down more than its counterpart 20 meters above it.

Einstein's General Theory of Relativity describes the effect on space and time measurements of an observer, either imbedded in a gravitational field or else undergoing non-uniform motion. The General Theory of Relativity also provides a theory of gravitation, which competes with Newton's theory of gravity. In addition to providing relativistic corrections to Newton's theory, Einstein's theory attempts to explain how the gravitational interaction arises. Newton's principle of action at a distance is replaced by a field concept. According to Einstein, matter warps the space around it. The warped space, in turn, affects the motion of the matter contained in it. The Sun, for example, warps the space in which the solar system is embedded, creating grooves in which the planets move. Determining the gravitational interaction of matter becomes a matter of geometry.

Before turning to the details of Einstein's model of gravity, we must first discuss the four-dimensional space-time continuum of relativity, which replaces Newton's concept of three-dimensional space and absolute time. In the Special Theory of Relativity, we discovered an intimate relationship between space and time. Time is no longer independent of the position or motion of an observer. Shortly after Einstein's Special Theory appeared in 1905, Minkowski proposed a four-dimensional space-time continuum with time playing the role of the fourth dimension. In this space, every event is described by four co-ordinates or numbers. Three of these co-ordinates, x, y, and z, describe the spatial location of the event while the fourth component, t, describes the time when the event occurs. These co-ordinates are defined with respect to a stationary frame of reference. With respect to some other frame of reference moving with a constant velocity, v, with respect to the original frame, another set of co-ordinates x', y', z' and t' are defined. Einstein showed that the co-ordinates x', y', z' and t' are related to those of x, y, z, and t by the following formulae when the velocity, v, is along the x-axis:

$$x = (x' + vt')/\sqrt{1 - v^2/c^2} \qquad y = y' \qquad z = z'$$

$$t = (t' + vx'/c^2)/\sqrt{1 - v^2/c^2}.$$

The details of these equations are not important for the purposes of our discussion so don't let their apparent complication disturb you if you are not mathematically inclined. The important thing to notice about these equations, known as the Lorentz transformation, is that both the position, x, and the time, t, in the stationary frame, depends on both x' and t' in the moving frame. In Newtonian physics, the time in the moving frame would be identical to the time in the stationary frame, i.e. $t = t'$ and $x = x' + vt$.

We see, in relativistic physics, that space and time are interwoven. The contraction of length and the slowing down of moving clocks that the observer in the stationary frame observes occurring in the moving frame, may be described as rotations in the four-dimensional space-time continuum. From the Minkowski point of view, an interval of time may be regarded as a length in the four-dimensional space-time continuum lying in the t-direction. The time interval is multiplied by c so that it has the same dimension as a length. Let us reconsider the example of the lightning striking the two ends of a train that we discussed in the beginning of the last chapter. From the point of view of the observer at rest, the lightning strikes the front of the train $vL_0/c^2$ seconds before it hits the end of the train, as can be verified from Fig. 14.1. The spatial length of the train, on the other hand, is contracted according to the stationary observer and appears to have the length $L_0\sqrt{1 - v^2/c^2}$. The separation of the two events, which to the moving observer was purely spatial, become both spatial and temporal to the stationary observer, demonstrating the equivalence of space and time. The observation of the stationary observer may be obtained from the moving observer's observations by rotating in the space-time continuum about an axis perpendicular to the x and t- axes so that a length, which is purely spatial in the primed co-ordinate system of the moving observer, becomes both spatial and temporal in the unprimed co-ordinates system of the stationary observer.

Another example of this rotation in the space-time continuum was encountered when we considered the time dilation of a moving clock. In the frame moving with the clock, the two events of the clock reading 8:00 a.m. and 9:00 a.m. are separated by a purely temporal length. In the stationary frame, however, the two events are separated by both a spatial and temporal duration, since the clock moves with respect to the stationary observer. Einstein utilized the Minkowski four-dimensional space-time continuum to formulate his theory of gravity. He acknowledged his debt to Murkowski referring to his contribution as follows: "Without it, the General Theory of Relativity would perhaps get no farther than its long clothes."

Although the space-time continuum is four-dimensional in the Special Theory of Relativity, its geometry is still basically Euclidean. Space is not warped or curved. The shortest distance between two points is a straight line, as is the trajectory of a light beam. In the General Theory of Relativity, however, the geometry of the space-time continuum is no longer Euclidean; space is curved. The curvature of space was demonstrated, theoretically, using the equivalence principle and, subsequently, verified experimentally, with the detection of the curvature of a beam of light passing close to the Sun during a solar eclipse. The shortest distance between two points in the vicinity of the Sun is not a straight line. The most basic axiom of Euclidean geometry, which Kant had elevated to an *a priori* truth, is ironically and empirically untrue in the case of light passing close to the Sun. One must describe the four-dimensional space-time continuum in terms of non-Euclidean geometry, which, fortunately for Einstein, had been worked out by Lobachevsky, Gauss and Riemann in the nineteenth century.

In non-Euclidean geometry, the ratio of the circumference of a circle to its diameter is no longer equal to $\pi$ as it is in Euclidean geometry. We can show that the ratio of the circumference to the diameter of a circle in a rotating frame of reference is less than $\pi$ and hence, invoking the equivalence principle demonstrates the need to describe the space-time continuum embedded in a gravitational field in terms of non-Euclidean geometry. Let us consider a disc, which is a perfect circle at rest with the circumference equal to $\pi$ times its diameter ($C = \pi D$). Let us consider the same disc rotating such that its outer edge has the velocity v with respect to an observer sitting at rest at the centre of the disc. To this observer, a length on the edge of the disc will appear contracted by the Lorentz–Fitzgerald factor $\sqrt{1 - v^2/c^2}$ because of the motion of the outer edge. This observer will measure the circumference to be $\pi D \sqrt{1 - v^2/c^2}$. The diameter of the circumference will not appear contracted, however, because it lies perpendicular to the motion of the disc. The ratio of the circumference to the diameter of a circle in this rotating frame of reference is not $\pi$ but $\pi \sqrt{1 - v^2/c^2}$. The geometry in this frame of reference is non-Euclidean, and since a frame in non-uniform motion is equivalent to a frame in which there is a gravitation field, we see that the geometry of the space-time continuum is also non-Euclidean. Space is curved in a gravitational field. The slowing down of a clock is also easily demonstrated in our rotating frame. A clock at the edge of the disc will appear to slow down to our observer at the centre by virtue of its velocity.

The curvature of space is difficult to comprehend in three dimensions, let alone four dimensions, as required in relativity theory. Let us consider a two-dimensional example to give ourselves a feeling for non-Euclidean geometry. We shall compare a flat, two-dimensional plane, described by Euclidean geometry, with the surface of a sphere. On the flat plane, the shortest distance between two points is a straight line and the ratio of the circumference to the diameter of a circle is $\pi$. The surface of a sphere is a curved, two-dimensional space, described by non-Euclidean geometry. The shortest distance between two points is not a straight line but a segment of a circle. This circle is called a great circle and the path between the two points is called the geodesic. Let us now consider a circle inscribed on the surface of the sphere. This circle is defined as the locus of all points equidistant from the centre of the circle. If we were to measure the circumference and the diameter of this circle, we would discover that the ratio of these quantities is less than $\pi$.

In formulating the gravitational interaction between masses, Einstein does not use the concept of one mass exerting a force upon another. Instead, he calculates the curvature of the space-time continuum due to the presence of matter. He then assumes that a mass will travel along the shortest possible path in this non-Euclidean, four-dimensional space. The path, geodesic or world line, along which the particle travels, is determined by the curvature of the space and the initial position and velocity of the particle. The world line of the Earth is not due to the force exerted on it by the Sun, but rather because this world line is the shortest possible path it can find in the curved space-time continuum surrounding the mass of Sun. The space around the Sun is curved or non-Euclidean because of its gravitational mass.

Einstein's model of the gravitational interaction is able to account for something, which Newton's model could never explain. Mercury, the closest planet to the Sun, orbits the Sun every 88 days in an elliptical orbit. The entire orbit of Mercury rotates at an extremely slow rate about the Sun. The distance of closest approach, the perihelion, of Mercury, advances 43 seconds of arc per century. Since there are 360 degrees in a circle, 60 minutes in a degree and 60 seconds a minute, it would take a little more than 3,000,000 years for the perihelion of Mercury to make one complete rotation about the Sun. This extremely small effect, nevertheless, cannot be explained in terms of Newton's theory of gravity. Astronomers tried to explain this effect in terms of the interaction of the other planets on Mercury. In fact, the existence of a planet lying between Mercury and the Sun, called Vulcan, was postulated to explain the advance of the perihelion of Mercury. This planet was never discovered and the anomalous behaviour of Mercury remained a mystery until Einstein's General Theory of Relativity, which was able to explain this effect in terms of the curvature of space.

Dicke has since shown that the advance for the perihelion of Mercury could also be explained if the shape of the Sun were not spherical. Since the shape of the Sun is not known, this is also a possible explanation. The experimental evidence to support the General Theory of Relativity is not as conclusive as the evidence supporting the Special Theory of Relativity. Experimental confirmation of Einstein's Special Theory occurs every day in a laboratory of the elementary particle physicists who explore the properties of elementary particles using high-energy accelerators. Experimental confirmation of general relativity has only been obtained in three instances:

1. The bending of light by the Sun.
2. The shift of spectra from the surface of a white-dwarf star, from the surface of the Sun and from different elevations near the surface of the Earth.
3. An explanation of the advance of the perihelion of Mercury.

Research in general relativity continues. Two areas of research that are being actively pursued in general relativity are the search for a unified field theory and the detection of gravity waves. After Einstein published his General Theory of Relativity, he pursued the problem of finding an underlying structure of the gravitational field and the electromagnetic field, which would unite them. In the Special Theory of Relativity, he was able to find the underlying structure, which united the electric and magnetic fields. In the General Theory of Relativity, he showed that gravity may be regarded as a property of the space-time continuum. He tried to show that the electromagnetic field could also be explained as a property of the space-time continuum so he could unite the gravitational and the electromagnetic fields. He never succeeded during his lifetime to solve this problem.

The idea of gravity waves was proposed in analogy to electromagnetic waves. If the electromagnetic field has waves in the form of light, it would seem that the gravitational waves should also exist. Electromagnetic waves are generated through the acceleration of charge, so, perhaps, the acceleration of matter would produce gravity waves. Theoreticians estimated that gravity waves would be very weak. The experimental detection of gravity waves has not been established. Experimental work in this area continues.

General Relativity also plays a role in the study of cosmology and dark matter and energy, which we will study in Chapter 25.

# Chapter 16

# Kuhn's Structure of Scientific Revolutions and the Impact of the Theory of Relativity

Einstein's Special and General Theories of Relativity caused a revolution in scientific thought, which affected a number of other fields as well. We will study the nature of these two revolutions and their influence on the various aspects of human thought. Thomas Kuhn (1972), in his book, *The Structure of Scientific Revolutions*, proposes a model to explain the nature of how a scientific revolution takes place. We shall review Kuhn's theory and examine the Einsteinian revolutions in terms of it.

Kuhn does not regard the history of science as the accumulation of "the facts, theories and methods collected in current texts". He believes this view of history arises from the tradition of teaching physics from textbooks in which the true historical processes are suppressed and science is presented as an accumulation of knowledge. Kuhn sees the history of science as a competition between different worldviews in which revolutions periodically occur whenever a worldview fails to accommodate new observations or new ways of looking at older observations. He claims that:

> The early developmental stages of most sciences have been characterized by continual competition between a number of distinct views of nature, each partially derived from, and all roughly compatible with, the dictates of scientific observation and method.

One body of beliefs wins out over the others because it provides a more satisfying description of nature. This usually occurs as a result of some success that the new theory achieves. This success becomes the model or paradigm for future scientific work, which Kuhn labels as normal science. The paradigm becomes a conceptual framework into which all of nature is forced to fit. The paradigm reduces the number of things the scientists must consider. They can take the foundation and assumption of their work for granted and can, therefore, exert their full effort in extending, articulating and generalizing the current paradigm of normal science.

Within the context of normal science, novelty is suppressed. No changes in scientific thinking takes place until new facts arise, which cannot be accommodated by the old paradigm. Out of the frustration of the failure to fit the new information into the old conceptual framework, a new picture emerges. The new paradigm is usually proposed by someone new to the field who has not become set in his or her ways through frequent use of the old paradigm. With the proposal of a new paradigm a revolution in thinking takes place in which the old view and new view are in conflict and competition. Eventually, one of the theories triumphs and a return to normal science ensues in which the whole revolutionary process may repeat itself. Kuhn does not regard only the major upheavals in thought such as those brought about by Copernicus, Newton or Einstein as the only scientific revolutions. He also regards the discovery of x-rays or Maxwell's formulation of electromagnetic theory as a scientific revolution as well. Each of them brought about a new way of thinking, a new framework for organizing information.

Perhaps the most controversial aspect of Kuhn's view of science history is his rather startling claim that the resolution of the conflict between the two competing theories during the revolutionary period is not really rational. The two sides of the controversy make different assumptions, speak a different language and hence, really don't communicate with each other. Max Planck, the man who kicked off the quantum revolution expressed this idea best:

> A new scientific truth does not triumph by convincing its opponents and making them see the light, but rather because its opponents eventually die, and a new generation grows up that is familiar with it.

An illustration of what Kuhn and Planck are saying is found in the response of William Magie to Einstein's Theory of Relativity. Magie felt that, although the formulae of Einstein described phenomena accurately from a mathematical point of view, they did not really explain the phenomena. Only an explanation, which he could comprehend with his common sense, i.e. fit into his notion of normal science, was valid as far as he was concerned. His attitude is reflected in the following passage from his presidential address to the American Physical Society in 1911:

> The elements of which the model is constructed must be of types, which are immediately perceived by everybody as the ultimate data of consciousness. It is only out of such elements that an explanation, in distinction from a mere barren set of formulae, can be constructed. A description of phenomena, in terms of four dimensions in space, would be unsatisfactory to me as an explanation, because by no stretch of my imagination, can I make myself believe in the reality of a fourth dimension. The description of phenomena in terms of a time, which is a function of the velocity of the body on which I reside, will be, I fear, equally unsatisfactory to me, because, try I ever so hard, I cannot make myself realize that such a time is conceivable.

In contrast to the response of Magie, a supporter of the old Newtonian paradigm, we have the euphoric response of Eddington. Eddington was a member of the 1919 solar eclipse expedition to Africa, which detected the bending of starlight by the sun. He finds Einstein's new ideas exhilarating and liberating as this excerpt from his early writings, reveals:

> To free our thought from the fetters of space and time is an aspiration of the poet and the mystic, viewed somewhat coldly by the scientist who has too good a reason to fear the confusion of loose ideas likely to ensue. If others have had a suspicion of the end to be desired, it has been left to Einstein to show the way to rid ourselves of those 'terrestrial adhesions to thought.' And in removing our fetters he leaves us, not (as might have been feared) vague generalities for the ecstatic contemplation of the mystic, but a precise scheme of world-structure to engage the mathematical physicist.

Kuhn maintains that a scientific revolution occurs when the existing forms can no longer accommodate new facts and offer no possibility of resolution. The only recourse is to a new system with a different set of values. It is obvious that Magie and Eddington have a different set of values and hence, could never convince each other which framework is the correct one. It can be argued that Newtonian physics is not really incompatible with Einsteinian physics since, as we saw earlier, that the relativistic effects disappear as the velocities involved become very small compared with the velocity of light. While it is true that the formulae become identical in the limit of small velocities, the interpretation of space and time still remains radically different for the Newtonian and the Einsteinian physicist.

Because of the fact that the new theory always incorporates the valid results of the old theory, science continues to make progress by definition. Kuhn claims that the progress made by science arises from the fact that the range of phenomena explained is continually increasing, not because the Einsteinian point of view is superior to the Newtonian point of view. There are no absolute truths for Kuhn. In this sense, his ideas have been influenced by Einstein's Theory of Relativity. Kuhn points out that scientific revolutions are often fomented by those who are new to the field and/or those who do not belong to the scientific establishment. No one could provide a better example of this viewpoint than Einstein as the following brief sketch of his life reveals.

Einstein was born in Bavaria in 1879 to a middle class Jewish family. He was a slow learner as a child. His parents feared that he was retarded because he was so late in learning how to speak. He was never a good student since he was given to daydreaming. At the age of fifteen, he dropped out of school and traveled in Italy. He finally settled down, applied for admission to university, and failed the entrance exam. He went back to secondary school for a year and passed the entrance exam to the Swiss Federal Polytechnic School in Zurich. He was not a model student in university, either. He passed his exams by cramming his friends' notes. He preferred reading on his own to attending lectures. He found the whole experience of higher education appalling as the following remarks penned some years later reflect:

> ... after I had passed the final examination, I found the consideration of any scientific problem distasteful to me for an entire year. It is little short of a miracle that modern methods of instruction have not already completely strangled the holy curiosity of inquiry, because what this delicate little plant needs most, apart from initial stimulation, is freedom; without that it is surely destroyed ... I believe that one could even deprive healthy beast of prey of its voraciousness, if one could force it with a whip to eat continuously whether it were hungry or not ...

Nevertheless, Einstein completed his Ph.D. in physics. Afterwards, he earned money as a tutor until an influential friend found him an undemanding job in the Swiss Patent Office in Berne. He held this position for seven years (1902–1909),during which he published his most famous papers. In 1905, the year his papers on the Special Theory of Relativity appeared, he also published two other major works, which, in themselves, also represented revolutions in physics thinking. It was for his 1905 paper on the photoelectric effect, which established the existence of the photon, and not for relativity that he received the Nobel Prize in 1921. In that same year of 1905, he also published his paper describing Brownian motion, which established the first direct detection of atoms. Three revolutionary papers in one year were produced and all the time while working in the patent office in Berne.

It was in 1909, after he had completed his most revolutionary work with the exception of the General Theory of Relativity that Einstein began to receive academic acclaim. In the years following 1909, he accepted professorial chairs from Prague, Zurich and Berlin. In 1932, as Hitler came to power in Germany, Einstein fled to the United States where he spent the remainder of his life at the Institute for Advanced Studies at Princeton. He was also a great humanist devoting much of his efforts to world peace.

Einstein's Theory of Relativity caused as large a response in the non-scientific world as it did in the scientific world. The overwhelming response of the layman was bewilderment because the notions of relativity violated their notion of common sense very much as it had violated the common sense of the physicist, Magie. Relativity for them was strictly a mathematical model, which they could not relate to intuitively. One aspect of Einstein's ideas, however, became very popular and that was the notion of relativity. If everything is relative in the physical world, it was argued then that the same must be true of things in the world of art, morals or ideas. Some people were appalled by this notion and were opposed to Einstein's ideas because they considered them downright immoral. Other thinkers found the aspect of Einstein's physical theories extremely liberating. For example, Jose Ortega y Gasset wrote in his book, The Modern Theme: "The theory of Einstein is a marvelous proof of the harmonious multiplicity of all possible points of view if the idea is extended to morals and aesthetics, we shall come to experience history and life in a new way. It is the same with nations. Instead of regarding non-European cultures as barbarous, we shall now begin to respect them, as methods of confronting the cosmos, which are equivalent to our own. There is a Chinese perspective, which is just as justified as the Western one.

Another thinker, Paul Squires, saw in Einsteinian relativity support for the notion of Gestalt psychology. He likens traditional psychology to Newtonian physics and Gestalt psychology to Einsteinian physics. He wrote,

> For the traditional psychology, particular experiences and bouts of behaviour possess a more or less absolute character. For the Gestalt psychologist, any aspect of mentality has meaning only in its relation to the larger context, the whole, in which it exists.

The art critic, Thomas Craven, in an article written in 1921, claimed that Einstein's ideas support the position of the modern artists. He wrote,

> While the celebrated physicist has been evolving his shocking theories of the course of natural phenomena, the world of art has suffered an equivalent heterodoxy with respect to its expressive media. This revolt has sprung from conviction that the old art is not necessarily infallible, and that equally significant achievements may be reached by new processes and by fresh sources of inspiration.

Craven regards the rigidity of "the old art" as corresponding to the "immovable reference-body" of Newtonian physics. He claims just as Einstein's theory changed Newton's concepts of absolute space and time "similarly has the modern painter broken the classical tradition." Rather than clinging to the rigid laws of photographic vision, he claims that the artist injects his own "personal feeling". Instead of arriving at an absolute truth "which serves all purposes of illustration" but "reveals nothing psychologically", artists achieve a greater truth, a psychological truth, relative to themselves.

## Chapter 17

# The Structure of the Atom

Early in the 1800's, when John Dalton proposed the atomic structure of matter; he considered the atom to be the simplest structure possible in nature, the ultimate subdivision of matter. The indivisibility and immutability of these ultimate building blocks of nature were implied in the very term, atom. The atom, as originally defined by the early Greeks, Leucippus and Democritus, were also thought to be indivisible and unchanging. Dalton had no notion that the atom was composed of the still smaller particles, the electron, the proton and the neutron. The phenomenon of electricity was not connected to the existence of electrons and protons or to any charged particle, for that matter. Rather, it was believed that electrical phenomena could be explained in terms of two different kinds of electric fluids, one positive and the other negative. The term current itself, used to describe a flow of electrically charged particles, retains the original connotation of the fluid nature of electricity. The electric fluid used to describe charge was considered to be massless and was not associated with matter or atoms. It was an independent constituent of nature.

Our description of electric and magnetic phenomena in Chapter 10 was not really historically accurate. The laws of electricity and magnetism were originally formulated in terms of the electric fluids rather than in terms of charged particles. All of the notions of electromagnetic phenomena, including Maxwell's equations, were conceived in terms of the density and the current of the positive and negative electric fluids. It was not until after the electromagnetic interaction was understood that it was realized that the density of the two electric fluids was the density of electrons and protons and that currents were generated by the movement of these particles. Just as the notion of caloric fluid disappeared once it was realized that heat was due to the

157

motion of the atom, so, too, the notion of electric fluids disappeared once it was realized that electric phenomena were due to the underlying electrical structure of atoms.

We now turn to a description of the experiments, which led to the discovery of the underlying electrical structure of the atom. Students are often baffled as to how the physicists know that an atom consists of a positively charged nucleus about which negatively charged electrons orbit. They become skeptical about the existence of the electron when they discover that the electron has not and cannot be visually observed. In order to counteract this skepticism, we shall carefully describe a series of experiments stretching over a period of almost 100 years in which the existence, charge and mass of the electron were established and in which the underlying structure of the atom was discovered.

The first hint of any connection between the atom and electrical phenomena came from the observation of the process of electrolysis. A typical electrolysis set up involves an anode or positive electrode and a cathode or the negative electrode, which are two metal rods connected to the positive and negative terminals of a battery and are immersed in water. A current is able to flow through the water because the water molecules, $H_2O$, separate chemically into two positively charged hydrogen ions, $H^+$ and one doubly negatively charged oxygen ion, $O^{2-}$. As the current flows in this circuit, hydrogen and oxygen gas will be attracted to and collect at the negative cathode and the positive anode respectively. The volume of hydrogen gas that collects is twice that of the oxygen gas and is proportional to the total amount of electricity that flows in the circuit. An ion is an atom, which carries a charge. Because there is an excess of negative charge at the cathode, the $H^+$ ions collect there, become neutralized and form hydrogen gas. The oxygen ions drawn by the positive charge collect at the anode, are neutralized and form oxygen gas. The ratio of the volume of hydrogen gas to oxygen gas is two to one because there are two hydrogen atoms to one oxygen atom in the water molecule.

Other electrolysis experiments showed that most atoms are capable of carrying a charge. One such experiment led to the process of silver plating. Let us consider the following set up where we choose for the anode, a pure silver rod, and for the cathode, a copper rod. Instead of water as our conducting medium, we choose a solution of silver nitrate, $AgNO_3$. When silver nitrate is dissolved in water, silver ions, $Ag^+$, are attracted to the copper cathode to which they attach themselves forming a silver plating on the cathode. The nitrate ions, $NO_3^-$, are attracted to the silver anode where they combine chemically with the silver to produce more silver nitrate. The level of silver nitrate remains the same as a result of this process, silver, however, is transferred from the silver anode to the copper cathode where it forms a uniform plate. One can reverse this process and remove the silver plate from the copper rod by changing the direction of the current. This is easily accomplished by switching the terminals of the battery.

The electrolysis experiments first performed by Humphrey Davy and Michael Faraday in the early part of the nineteenth century were motivated by a desire to study the passage of an electric current through a liquid, but they also revealed the existence of ions, i.e. atoms carrying a charge. Faraday, during this period, also began to study the problem of the passage of an electric current through a gas. He discovered that gases are excellent insulators and that the conductivity of the gas increases as its density decreases. Faraday's research was limited by the primitive state of vacuum techniques, at the time. It was not until 1854 when Heinrich Geissler invented a vacuum pump, which enabled him to evacuate a glass tube that the passage of an electric current through a gas could be properly studied. The cathode and anode are attached respectively to the negative and positive terminal of a high voltage battery creating a strong electric field inside the discharge tube. The passage of an electric current occurs as a result of the ionization of the gas molecules (or atoms), the process whereby an electron is removed from the gas molecule (or atom) leaving a positively charged ion. At the time these experiments were first performed, it was thought that the ionization was due to the strength of the electric field in the tube, which was thought to tear apart the molecules of the gas. We have since discovered that the ionization is due to the effects of cosmic radiation, the high-energy charged particles, x-rays and gamma rays that are continually streaming towards earth from outer space.

Once the molecule is ionized, the positive (negative) ion begins to accelerate towards the anode (cathode), each gaining kinetic energy from the electric field. If the gas is dense, they will soon collide with a neutral molecule, lose their kinetic energy, and begin accelerating once again toward their respective electrode. In this way, a small current is able to pass through a dense gas. As the tube is evacuated so that the density of the gas becomes significantly lower, the charge particles do not collide as often with the neutral molecules of the gas. They, therefore, have time to build up more kinetic energy so that when they finally collide with a neutral molecule, they can ionize it creating another electron-ion pair. The new ion pair will then go on to create many more additional ion pairs. This process snowballs and an electric discharge between the cathode and the anode takes place, an appreciable current begins to flow, and the tube begins to glow visibly. Not all of the electrons and ions in the gas actually reach the anode or the cathode. Some of them recombine on the way with new partners. Because of the kinetic energy the ion pair has acquired from the electric field, the neutral molecule formed through the recombination of the negative and positive ions is often in an excited state. It rids itself of this excess energy by radiating visible light, which partly accounts for the glow from the center of the tube.

There is also a glowing of the glass at either end of the tube, which is due to the stream of charged particles. When this glow was first observed, it was not understood what caused it. It was thought to be some kind of radiation. The glow at the anode caused by the electrons, for example, was attributed to cathode rays, which presumably emanated from the cathode. By cutting a hole in the electrodes, William Crookes was able to study the nature of the radiation causing the glass of the tube to glow. William Crookes showed that the location of the glow could be displaced by a magnet. He cited this as evidence that the so-called radiation was actually due to a stream of charged particles.

This work was corroborated and further developed by J.J. Thomson in 1897, who showed that the stream of particles could also be deflected by an electric field. By comparing the deflections due to the electric and magnetic fields, Thomson was able to determine the ratio of charge to mass for the charged particles. His results with the positive ions were in agreement with the results of other determinations of the charge to mass ratio of heavy ions obtained with electrolysis studies. His results with the negatively charged particles were extremely surprising. He found that the ratio of charge to mass of these negatively charged particles was thousands of times larger than the same ratio of the positive ions observed in electrolysis. If one assumed that the magnitude of these newly discovered particle's charge was the same as that of the hydrogen ion, then, its mass would be only 1/2000 the mass of the hydrogen atom. Thomson associated this particle with the electron, named in 1891 by George Stone, whose existence was postulated as the ultimate particle of electricity. Thomson's result established the electrical composition of the atom.

Soon after Thomson's discovery of the electron, two models of the atom arose. According to the proponents of one point of view, the atom was a miniature solar system. There was a positively charged nucleus at the centre of the atom around which orbited the negatively charged electrons, as pictured in Fig. 17.1. This model, which eventually developed into our present-day notion of the atom, was in fierce competition with a model proposed by Thomson himself, which he called "the plum pudding model" of the atom. He assumed that the atom was a sphere of positive charge in which the electrons were embedded like plums in a pudding as pictured in Fig. 17.2. In both models, the atom was basically neutral. The atom became a positive or negative ion as electrons were either lost or gained. Both models explained the results of the electrolysis and gas discharge experiments.

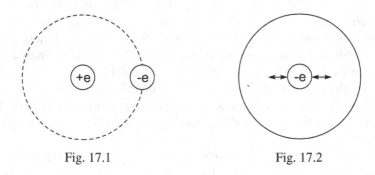

Fig. 17.1                    Fig. 17.2

The plum pudding atom, however, gained favour over the miniature solar system atom because it provided a more satisfactory description of the electromagnetic radiation emitted by matter. The light emitted by various materials was examined by passing it through a prism before it fell upon a photographic plate. This enabled the experimentalists to determine the spectral distribution of the light being radiated by a particular material. They found that the spectra from the closely-packed solids and liquids are continuous, but, that the spectra from the gases are distinguished by discreet lines. Apparently, the continuous spectra were due to the external motion of the atom as a whole, due to heat. This was verified by the fact that the distribution of frequencies changed with the temperature, so that the higher the temperature the higher the frequency of the radiation. This is to be expected since the frequency of the light emitted, according to Maxwell's theory of radiation, is related to the frequency of the charged particles emitting the radiation. As the solid or liquid is heated, the atoms move faster and, hence, one expects a shift to higher frequencies. Both the plum pudding and solar system models of the atom could account for the continuous spectrum of radiation. The spectra from the gases were not as easily explained. Each gas had its own characteristic set of discreet lines implying that this radiation was due to internal motion within the atom. This hypothesis was confirmed by the fact that the heating of the gas did not change these characteristic frequencies. They were unaffected by the external motion of the atom and, hence, were due to the internal motion of the atom. The plum pudding model explained these discreet lines as due to the oscillations of the electrons in the pudding of positive charge. Proponents of the solar system model tried to explain these frequencies in terms of the frequency of the electron's orbit about the nucleus. According to Maxwell's theory, however, a charged particle moving in a circular orbit is constantly accelerating and, hence, would be continuously radiating light. As the electron continued to radiate light, it would lose energy, and, hence, fall into an orbit closer to the nucleus where it would lose more energy and so on until it spiraled into the nucleus. This defect of the solar system model of the atom, known as the "spiral death", caused it to fall into disfavour. The solar system atom would arise again and triumph but, for the time being, the plum pudding model provided the best description then of the atom.

Although Thomson's gas discharge experiments had established the existence of the electron, the charge of the electron had never been measured nor had the hypothesis that the electron was the ultimate unit of charge been verified. It wasn't until 1909 that Robert Millikan completed the work begun twelve years earlier by Thomson with his famous oil drop experiment. Millikan sprayed oil drops between two metal plates through the narrow nozzle of an atomizer. Many of the oil drops became charged in this process losing or gaining electrons through the act of subdividing or by rubbing against the inner wall of the nozzle or each other. After spraying the oil drops between the two metal plates, he would apply a known electric field between the plates, which would accelerate the oil drops upwards. By measuring the rate of the upward acceleration, Millikan was able to determine the ratio of charge to mass on each oil drop. He would then turn off the electric field and observe the oil drops falling under the influence of gravity. By measuring their terminal velocity, he was able to determine the mass of the oil drop. Once he knew the mass of the oil drop he could then determine the absolute value of the charge it was carrying. He found that the charge on the oil drops came in discreet units, which he labeled e. This unit, e, is the absolute charge of the electron and is a fundamental constant of nature. It is an extremely small number equal to $1.6 \times 10^{-19}$ coulombs. A coulomb is the amount of charge that flows through a standard 100-watt light bulb in approximately one second. The smallest charge observed for an oil drop was –e. The charges on the other oil drops were always equal to an integer times e. Thus, one observed oil drops with charge +e, –e, +17e, –10e, +3e, etc., but never with charges like 0.5e, –1.1e or –2.37e. Millikan concluded that an electron carries the charge –e and that the charge on his oil drops was due to an excess or lack of a given number of electrons. The oil drop with charge +17e was missing 17 electrons whereas the oil drop with charge –10e had 10 extra electrons.

Once the charge of the electron was known, one was able to determine the mass of the electron using Thomson's determination of the charge to mass ratio. One discovers, in this way, that the electron is an enormously small object with a mass of only $9.1 \times 10^{-28}$ grams. Exploiting Thomson's determination of the charge to mass ratio of the hydrogen ion, $H^+$, and assuming that it has the charge +e, we find that the mass of the hydrogen atom is $1.66 \times 10^{-24}$ grams or almost 2000 times the mass of the electron. Once the mass of the hydrogen atom is determined, the absolute masses of all the other atoms are determined also since the relative mass of the atoms is known from the work of the chemists of the first half of the nineteenth century. The oxygen atom is approximately 16 times the mass of the hydrogen atom and the carbon atom is 12 times the mass of the hydrogen atom. From the absolute weights of the atoms, we can determine the number of hydrogen, carbon or oxygen atoms in one gram of hydrogen, 12 grams of carbon or 16 grams of oxygen, respectively. This number, which is called Avogadro's number, is equal to $6 \times 10^{23}$. Making use of our knowledge of the density of graphite, which is pure carbon, we can determine the volume occupied by a carbon atom and, hence, its radius, which turns out to be approximately $10^{-8}$ cm.

Actually, Avogadro's number was not originally determined by Millikan's experiment of 1909 but by Einstein's analysis of Brownian motion in 1905. Brownian motion, first observed in 1827 by the botanist Robert Brown, is the observed erratic, jerky motion of microscopic pollen grains suspended in water. The smaller the pollen grain, the more violent is the motion of the grain. Einstein pointed out in 1905 that this jerky motion is due to the fact that the pollen grain is being constantly bombarded by water molecules. From his detailed calculations of the pollen grain's motion, Einstein was able to determine the absolute weight of various atoms and, hence also Avogadro's number. His analysis of Brownian motion actually provided the first direct detection of the atom.

Additional evidence for the composite nature of the atom was gathered towards the end of the nineteenth century as new forms of radiation were discovered. Roentgen discovered x-rays in 1895 when working with his gas discharge tube, he noticed a glow in some fluorescent material lying near his apparatus. These rays were produced as a result of the bombardment of the atoms of the anode by electrons being accelerated there by the electric field in the gas discharge tube. As a result of the bombardment by the electrons, the atoms in the anode were excited and, as a consequence, emitted x-rays. It was not until the Van Laue experiments of 1912 that it was realized that x-rays are electromagnetic radiation of extremely small wavelength. The wavelengths of x-rays are of the order of $10^{-8}$ cm, the distance between atoms, which explains their ability to penetrate solid matter. Van Laue passed an x-ray beam through a crystal in which the atoms are arranged in an orderly array. He found that the x-rays displayed the characteristic diffraction pattern one observes for light passing through a system of slits. The spaces between the atoms in the crystal served as the system of slits. From the diffraction pattern and the knowledge of the spacing between the atoms in the crystal, other investigators were able to determine the wavelength of the x-ray. It was discovered that the x-rays produced by a particular atom had certain discreet characteristic frequencies analogous to the situation with the discreet lines of visible light emitted by gas atoms.

Roentgen's discovery of x-rays stimulated other researchers to look for still different forms of radiation. In 1896, Becquerel began a series of experiments with phosphorescent materials such as the uranium compound, potassium uranyl sulfate, which glows in the dark after it is exposed to sunlight. Becquerel was hoping to show that the radiation from phosphorescent materials contained x-rays. Quite by accident, he made another discovery, which turned out to be far more important. Becquerel left a sample of the potassium uranyl sulfate in his desk drawer with some photographic film and discovered, much to his surprise, that, although no light had fallen upon the film, it was still exposed. Apparently, potassium uranyl sulfate emits radiation on its own without being exposed to the sunlight. Becquerel isolated the source of the spontaneous of radiation and found that it was due to uranium. Becquerel's effort led to the work of Pierre and Marie Curie who went on to investigate the nature of radioactivity (the spontaneous emission of radiation from matter, a term coined by Marie Curie) and to isolate a number of radioactive elements such as polonium, thorium and radium.

Subsequent investigation revealed that radioactive atoms emitted three forms of radiation, which were labeled alpha rays, beta rays and gamma rays. Beta rays were identified by Becquerel as high-energy electrons. The gamma ray was discovered by P. Villard and identified as electromagnetic radiation with a wavelength even shorter than that of x-rays. Alpha rays were discovered by the Curies and Earnest Rutherford and were identified as a stream of particles with the charge, +2e, and a mass four times that of the hydrogen atom. In other words, the alpha particle is nothing more than the nucleus of the helium atom without its two electrons. The ultimate source of radioactivity within the atom is the nucleus and we shall, therefore, defer our discussion of radioactivity until we come to our treatment of nuclear physics. The importance of radioactivity for the study of atomic physics is twofold. Firstly, it provides, in itself, additional evidence for the composite nature of the atom. Secondly, and perhaps more importantly, Rutherford realized that, because of its charge, alpha rays could be used as a tool to probe the structure of the atom. He and his colleagues, Marsden and Geiger, devised a scattering experiment in which a collimated beam of alpha particles was directed at a thin gold foil target. A schematic sketch of the experimental setup is shown in Fig. 17.3. The alpha rays originated at the source and pass through a slit in a thick sheet of lead in order to produce a collimated beam. The collimated beam then passes through the gold foil where it is scattered by the individual gold atoms. The scattered alpha rays are then detected with a fluorescent screen to which is attached an eyepiece. Whenever an alpha particle strikes the fluorescent screen, a spark of light is produced, which is visible through the eyepiece. The fluorescent screen and eyepiece can be moved about the circle so that the number of alpha particles scattered at each angle can be counted.

When Rutherford and his colleagues performed their experiment, they believed that they would establish the experimental validity of J.J. Thomson's plum pudding picture of the atom in which a number of electrons floated in a spherical blob of positive charge with a radius of approximately $10^{-8}$ cm. They, therefore, expected that the alpha particle would pass through the atom without scattering very much because the electrons were too light to affect the alpha particles very much and because the positive charge was spread out over the whole atom. The alpha particle was expected to be only nominally scattered from its original path.

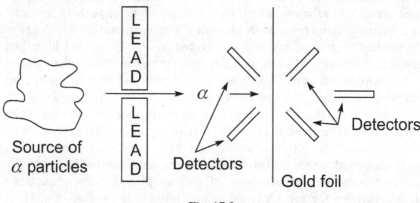

Fig. 17.3

Imagine what a surprise it must have been for Rutherford and his co-workers when they discovered that a number of alpha particles suffered considerable deflections and, in fact, some alpha particles had actually been scattered completely into the backwards direction, i.e., scattered back into the direction from which they originally came. The observed scattering of the alpha particles could only have been caused by the positive charge of the nucleus since the electrons do not have enough mass to scatter the alpha rays very effectively. But, if the positive charge of the nucleus were responsible for the large angle scattering of the alpha particles, then, the positive charge had to be concentrated into a considerably smaller space as is illustrated in Fig. 17.1. In fact, Rutherford showed that the positive charge had to be concentrated in a nucleus with a radius of less than $10^{-12}$ cm. The picture of the atom completely changed overnight. The miniature solar system atom came back into favour. The atom now consisted of a nucleus, which contains all of the positive charge in a sphere less than $10^{-12}$ cm in radius. (It is now known that the radii of nuclei vary from 1 to $7 \times 10^{-13}$ cm.) The electron circles the nucleus with an orbit whose radius is of the order of

$10^{-8}$ cm. Most of the atom therefore consists of empty space. This was the only picture of the atom consistent with the result of Rutherford's experiment.

This picture of the atom presented a rather grave problem, however. According to the classical theory of electromagnetic theory, any particle accelerating will radiate light energy. Since the electron orbiting the nucleus is continually accelerating, it should also be continually radiating light and hence, losing energy. As it loses energy, it spirals into the nucleus until it is finally absorbed. Essentially, the miniature solar system atom is unstable from the point of view of classical electromagnetic theory. The solution of this problem, which Bohr proposed two years later in 1913, is related to the quantization of the energy of light discovered earlier through the work of Max Planck in 1900 and Albert Einstein in 1905. Before continuing with our story of the atom, we turn our attention to the discovery of the photon and the quantization of energy.

# Chapter 18

# The Quantization of Energy

The exploration of the atom at the turn of the century provided the physicists of the day with a never-ending stream of surprises and shocks as one new discovery followed another. The most disturbing of all the new results of this period, however, occurred as a result of Max Planck's study of the energy distribution of electromagnetic radiation emitted from the surface of a hot body. According to Maxwell's theory, an electric charge, when accelerated, emits electromagnetic radiation. As a body is heated, the atoms composing it begin to oscillate rapidly, which causes them to emit radiation. In order to study the distribution of energy with respect to the frequency of this thermally induced radiation, a black body radiation device was constructed as depicted in Fig. 18.1. It consisted of a hollow spherical shell of metal in which a small hole was drilled.

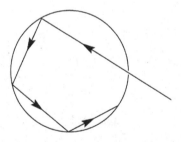

Fig. 18.1

The lines in the figure represent a beam of light entering the sphere, bouncing off the inner walls of the sphere and eventually being absorbed. Since it is difficult for light to leave the sphere once entering the interior of the hollow shell, the radiation inside the shell is due almost entirely to the thermal emission of electromagnetic radiation from the inner walls of the cavity. Any light, which does enter the cavity, is quickly absorbed by the inner walls. The distribution of energy of the thermal radiation can be studied by observing the radiation emerging from the hole in the shell. The results of this study are shown in Fig. 18.2 where the intensity of radiation is plotted versus the frequency for two different temperatures. The total amount of energy radiated increases rapidly with the temperature. It is proportional to the fourth power of the temperature. (If E is the total energy, T, the temperature, and k, a constant, then $E = kT^4$). As the temperature increases, the distribution shifts to the higher frequencies. The frequency for which the intensity is a maximum is proportional to the temperature.

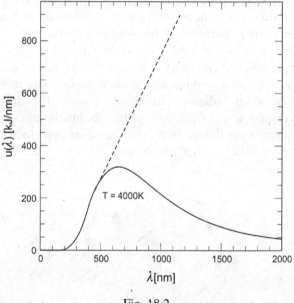

Fig. 18.2

The experimentally observed energy distribution represented by the solid curve differed very radically from the theoretical prediction depicted by the dashed curve in Fig. 18.2. It was believed that the higher frequencies would dominate the distribution since the probability of exciting a frequency was thought to be greater, the higher the frequency. This prediction was not fulfilled, however, and the intensity, instead of steadily increasing with frequency, reached a maximum value and then decreased to zero. This result was a total mystery to the physicists of the day. No one was able to explain why the intensity did not continue to increase with frequency.

Max Planck devoted a great deal of effort trying to derive the experimentally observed distribution from a strictly theoretical point of view. Not meeting with success, he attacked the problem from a phenomenological point of view. Rather than trying to derive the correct formula from first principle, he first searched for an algebraic formula, which would just describe the experimental data. Using trial and error and guided by previous attempts at a solution, Planck finally found a formula, which represented the data. Working backwards from the answer, Planck then searched for a way to derive his formula. He discovered he could obtain his formula if he made the assumption that the radiation of frequency, f, was absorbed and emitted in bundles of energy equal to hf where h is a constant, called Planck's constant, and is an extremely small number. The bundles of energies are called quanta (quantum is the singular) or photons.

The implication of Planck's result is that the energy of light is not continuous but is packaged in discreet bundles of energy called quanta or photons. According to Planck's hypothesis, a frequency, f, cannot be excited unless an amount of energy, hf, is provided to create a photon. Planck's constant, h, is extremely small and, hence, the amount of energy to excite a particular frequency is not much, however, it takes more energy to create the higher frequency photons than the lower frequency photons, and, hence, the probability of exciting the higher frequencies is not always greater than that of exciting the lower frequencies. Also, there is a cutoff of frequencies in Planck's model so that the infinite number of high frequencies can no longer contribute to the black body radiation. If the total amount of energy available for thermal radiation is $E_o$, then, the frequency $f_o$, such that $E_o = hf_o$ is the highest possible frequency that can contribute. There just would not be enough energy available to create a photon with frequency greater than $f_o$.

Planck's hypothesis that the energy of light is quantized explained the distribution of energy of black body radiation. Because Planck's constant is so small and, hence, each quantum or photon actually carries such a small amount of energy, Planck's hypothesis does not conflict with the experimentally observed continuity of light. Light rays are composed of literally millions and millions of tiny bundles of energy or photons, each of which carry a minuscule amount of energy. The amount of energy carried by the most energetic photon conceivable, for example, a photon emitted by a radioactive nucleus whose frequency corresponds to the gamma ray range, is equal to the amount of kinetic energy a drop of water would acquire falling under the influence of the earth's gravity the distance of $10^{-6}$ cm, the height of 100 atoms.

Although Planck's quantization of the energy of light did not violate any of the empirically observed facts of nature, the idea of the quantum violated the notions of classical physics held by his contemporaries. According to their way of thinking, most quantities including energy are continuous.

The first break they had known in this tradition of continuity was the discovery of the atom, which revealed that matter was discontinuous. With the discovery of the electron, the discontinuity or quantization of charge was also revealed. Although these concepts encountered some initial resistance, they could be incorporated into the fabric of classical physics because these discontinuities could be associated with the existence of particles, which were always regarded as discreet. The discontinuity of the energy of light or electromagnetic waves was inconceivable, however. Energy, first of all, was always considered a continuous quantity, even for discreet particles. But what made Planck's proposal even more mysterious was that it was associated with light, which ever since the diffraction experiments of Young, was considered to be a wave and, hence, continuous. Planck realized that his quantum hypothesis was in contradiction with the classical physics of his day. Referring to his work, he remarked to his son in Berlin in 1900, "Today, I have made a discovery as important as that of Newton."

The full implication of Planck's idea was not understood for five years until Einstein exploited the quantum hypothesis in order to explain a new experimental result know as the photoelectric effect. The photoelectric effect was first discovered by Hertz as early as 1887. It is the effect, which has since been exploited in a number of devices such as the "electric eye" and the photographer's light meter. The effect consists of the observation that, when light of a sufficiently high frequency falls upon a metallic surface, electrons (referred to as photoelectrons) are ejected from the metal. This emission of photoelectrons from the metal can be understood from a classical point of view. The light, which falls upon the metal is absorbed by the electrons inside. After a sufficient amount of time passes, the electron absorbs enough energy to overcome the electromagnetic forces, which holds it captive in the metal and it is then free to leave the metal as a photoelectron.

There are three aspects of this photoelectric effect, however, which cannot be understood in terms of classical physics. The first is the threshold effect of frequency. Unless the frequency of the light is above the threshold frequency, no photoelectrons are ejected from the metal, no matter how high the intensity of the light. There is no reason why this effect should depend upon the frequency, as the experimental data seemed to indicate. The second peculiar aspect of the photoelectric effect is that photoelectrons appear the instant the metal is irradiated by light. The electrons are ejected within $10^{-8}$ seconds of the light striking the surface of the metal, independent of the intensity of the light, as long as the frequency is high enough. This contradicts the classical picture of the electron absorbing energy from the light wave, since it is easy to show that $10^{-8}$ sec does not allow sufficient time for the electron to absorb enough energy from a wave to overcome the electromagnetic forces, which hold it captive within the metal. The third peculiar aspect of the photoelectric effect was pointed out by Leonard in 1902. He observed that, when the intensity of light was increased by moving the source closer to the metal, for example, the energy of the electrons ejected from the metal did not increase but rather more electrons appeared. If one increases the frequency of the light projected on the metal, however, then the energy of the ejected electrons will increase.

Einstein was able to explain all three of these mysterious effects in his brilliant paper of 1905, on the photoelectric effect for which he was awarded the Nobel Prize in 1921 (Einstein never received a Nobel Prize for his work on relativity). Einstein expanded Planck's quantum hypothesis by assuming that, not only the light induced by thermal radiation is quantized, but, that all electromagnetic radiation comes in bundles of energy or photons. The energy of an individual photon composing a light ray equals hf where h is Planck's constant and f is the frequency of the radiation. According to Einstein's hypothesis, at certain times, light behaves like a beam of particles where each particle is a photon. But, how does this explain, one might ask, the mysterious aspects of the photoelectric effect? Well, in order for an electron to escape the metal, it must have a sufficient amount of energy to overcome the normal electromagnetic forces, which keep it captive inside the metal, i.e. it must have enough energy to overcome the binding energy. If the frequency of the light is too small, then the energy of its quanta or photons is less than the binding energy and the photon cannot transfer enough energy to the electron for it to overcome the forces keeping it prisoner in the metal. If one increases the frequency, f, such that hf is greater than the binding energy, then, the photons can deliver enough energy to the electrons to allow them to escape. This explains the threshold effect.

The photon model also explains the instantaneous appearance of the electrons once the threshold frequency is surpassed. The electron does not absorb the required energy for ejection from the metal accumatively from a wave, but, rather, all at once from a single photon. The electron is ejected because it suffers a collision with a particle of light. Therefore, as soon as the collision occurs, the escape of a photoelectron is possible which is why electrons appear immediately after the metal has been irradiated by a ray of light, which is nothing more than a beam of photons.

The quantum hypothesis also explains why increasing the intensity does not increase the energy of an individual photoelectron. The energy of an ejected electron depends only on the frequency of the single photon, which knocks it out of the metal. Increasing the intensity of the light does not change the frequency of the photons, it only increases the number of them. This is why increasing the intensity increases the number of photoelectrons without increasing the energy of individual photoelectrons. The fact that the energy of the photoelectron depends on the frequency of the photon explains why the energy of the ejected electrons increase as the frequency increases.

Thus, Einstein was able to account for all the observationally known facts concerning the photoelectric effect by assuming the particle-like behaviour of light. Not only did he explain all of the observations known at the time he wrote his paper in 1905, but he also made exact mathematical predictions relating the energy of the ejected photo-electrons to the frequency of the light inducing the effect. Let W represent the binding energy and $f_0$ the threshold frequency at which the photoelectric effects first occur. Then the energy of the photons of frequency fo is hfo and equals the binding energy W. (In terms of an equation, we have $W = hf_0$ or $f_0 = W/h$). The energy of the escaped electrons ejected by photons with frequency f greater than $f_0$ is equal to the energy hf imparted by the photon minus the binding energy, W. Defining E as the energy of the ejected electron and using an equation once again, we have $E = hf - W$ for the photoelectron.

This precise mathematical prediction, made by Einstein in 1905, was not verified until eleven years later in a series of experiments performed by R.A. Millikan (the man who measured the absolute charge of an electron using oil drops). He verified that the Einstein formula was correct and that the energy of the photoelectrons was proportional to the frequency. The constant of proportionality is just Planck's constant, h, which Millikan found from his experimental results had the same value that Planck used when he explained black body radiation. Millikan's results confirmed, with great mathematical precision, that the energy of light is quantized and that light displays particle-like behaviour in the photoelectric effect.

Another experiment, which followed Millikan's photoelectric work also revealed the particle-like behaviour of light. This was the experiment of Compton in 1922 in which he scattered x-rays using electrons as the target. When a beam of x-rays is directed at an amorphous (non-crystalline) solid like graphite, most of the x-rays pass through the solid unscattered. Some of the x-rays are scattered in all directions by collisions with the electrons in the solid. Compton noted an effect, which bears his name, namely those x-rays, which are scattered by the electrons have a slightly smaller frequency after the collision and that the greater the angle of scatter, the more the diminution of the frequency.

The Compton effect is easily understood if the beam of x-rays is treated as a beam of photons. Each photon carries energy and momentum related to its frequency, f. The energy of the photon is equal to hf and its momentum to hf/c where c is the velocity of light. The momentum of a photon is related to energy by the theory of relativity and is equal to its energy divided by c. Although the electrons in the graphite are bound to the nucleus, they may be treated as unbound because the energy of binding is so small compared to the energy of the x-ray photons. In analyzing his experiment, Compton treated the electron and photon as two particles colliding with each other. Using this assumption, he was able to explain all of the details of the Compton effect. Once again, electromagnetic waves behave as though they are a beam of particles.

Newton's corpuscular theory of light seems to be valid in certain circumstances, namely for black body radiation, the photoelectric effect and the Compton effect. For other situations, however, such as diffraction, interference and refraction, light behaves as a wave. The dual nature of light provided physicists with the deepest mystery they had yet experienced. Light seemed to possess contradictory aspects. The dilemma was expressed succinctly by William Bragg, who had worked on the x-ray diffraction experiments, which had shown x-rays behave like waves. He suggested that on Mondays, Wednesdays, and Fridays, physicists believe and act on the basis of wave theory; but on Tuesdays, Thursdays and Saturdays, they follow the particle theory of light. To which some pundit added, "And on Sundays, they just pray for a resolution."

The prayed-for resolution was the modern theory of quantum mechanics that did not evolve until a number of equally nasty paradoxes were unearthed. The long and tortuous road to that resolution of the particle-wave duality of light began with the work of Niels Bohr.

# Chapter 19

# Bohr's Atom

In 1911 a young man, named Niels Bohr, journeyed from his native Denmark to England to study physics. The following year he joined the great Rutherford in Manchester and began to work on the problem of the atom. Bohr was a keen supporter of the Rutherford solar system model of the atom despite its problematic nature. Rutherford's model of the atom, in which the positive charge is concentrated in a nucleus whose radius is $10^{-13}$ cm around which electrons orbit with radii approximately $10^{-8}$ cm, was in deep conflict with classical electromagnetic theory on two accounts. According to electromagnetic theory an electron orbiting a nucleus should emit a constant stream of continuous electromagnetic radiation and because of the ensuing loss of energy quickly spiral in to the nucleus. The apparent stability of the atom as well as the presence of discreet lines in the energy spectrum of atomic radiation had to be explained within the framework of Rutherford's model. It was this problem that Bohr set out to solve in 1912.

In addition to explaining the stability of the atom and the discreteness of its radiation spectrum, Bohr also had the task of explaining certain regularities of the spectral lines that had been observed by the spectroscopists. As early as 1885, Balmer had shown that the observed frequencies of the hydrogen atom could be represented by the following simple mathematical formula, $f = R_y/h\{1/n^2 - 1/p^2\}$, where f is the frequency of the radiation, $R_y$ is Rydberg's constant equal to $3.27 \times 10^{15}$ sec$^{-1}$, h is Planck's constant and n and p are integers almost always less than 10 or so. Balmer's formula provided a very tidy description of the spectral lines observed in 1885. As more and more data was collected it was found that the spectral lines of hydrogen occurred at precisely those places predicted by Balmer's formula.

In 1908, Ritz discovered a regularity in the spectrum of light emitted by atoms, other than just hydrogen. He found that if $f_1$ and $f_2$ are the

177

frequencies emitted by an atom, then the same atom is very likely to emit the frequency $f_3 = f_1 + f_2$. This rule known as "Ritz's Combination Principle" follows trivially for hydrogen from Balmer's formulae. The novelty of his discovery, however, is that his principle applies to atoms other than hydrogen.

The work of Balmer and Ritz played an important role in Bohr's formulation of his atomic model by helping him make certain guesses he might not have made otherwise. The quantization of energy proposed by Planck and Einstein played an equally important role in Bohr's thinking. He was the first to apply the quantum concept to the atom. He assumed that the electron orbits the nucleus only along certain fixed elliptical orbits, each of which represents a different quantum state. He also assumed, in direct contradiction with electromagnetic theory that the electron would not radiate while moving along any of the allowed orbits. The electron is more or less stable in one of the allowed orbits or quantum states. Bohr did postulate, however, that the electron could jump to a lower energy orbit closer to the nucleus by radiating a quantum of energy. The electron would continue jumping to lower energy orbits emitting photons until it landed in the lowest energy orbit or ground state in which it could remain without ever emitting any more photons.

The energy and hence the frequency of the photon emitted by the electron as it jumps from an orbit with energy $E_2$ to an orbit with energy $E_1$ is related to the energy difference of the two orbits, $E_2 - E_1$. The relation between the frequency, f, of the emitted photon, the energy difference, $E_2 - E_1$ and Planck's constant, h, known as Bohr's frequency condition, is similar to Planck's original quantum condition, namely $hf = E_2 - E_1$.

According to Bohr's hypothesis, an electron not only may jump from an outer orbit to an inner orbit by emitting a quantum of energy, but it may also pass back from an inner orbit to a more energetic outer orbit by absorbing a quantum of energy. In order to make this transition it must absorb a quantum of energy exactly equal to the energy difference of the two orbits, $E_2 - E_1$.

Bohr's model of the atom violates the laws of classical mechanics and electromagnetism in a number of ways. First of all, the electron in the ground state is constantly undergoing acceleration without ever radiating light in direct contradiction with Maxwell's laws. Secondly, the frequency of radiation emitted by an electron as it jumps from one orbit to another is not equal to the frequency with which it orbits the nucleus. In classical electromagnetic theory, on the other hand, the frequency of the periodic motion and the frequency of the subsequent electromagnetic radiation are identical. Finally, in classical mechanics, an electron must orbit the nucleus in an infinite number of paths, differing by only an infinitesimal amount of energy. In Bohr's scheme, however, the number of orbits is severely limited by restricting the allowed orbits to those for which the angular momentum is equal to an integer times Planck's constant, h, divided by $2\pi$. If we represent the angular momentum by L, then $L = \ell \, h/2\pi$ where $\ell$ is an integer. The angular momentum of the electron is equal approximately to the product of its momentum times the radius of its orbit. This definition is exact if the orbit is a perfect circle.

By placing this restriction on the angular momentum, Bohr was able to obtain Balmer's formulae for the radiated frequencies of the hydrogen atom. Bohr was also able to calculate Rydberg's constant, $R_y$ and showed that it is simply related to the mass of the electron, $m_e$, the charge of the electron, e and Planck's constant, h, by the formula
$R_y = 2 \, \pi^2 \, m_e \, e^4 \, /h^3$. This result, in which one of the fundamental constants of nature was related to the others, was a great success and insured the acceptance of Bohr's model.

This model not only explained Balmer's formula for the hydrogen atom but it also explained Ritz's combination principle. Let us label the quantum states or energy levels of the atom by $E_1$, $E_2$, ... , $E_n$ where $E_1$ is the energy of the ground state, $E_2$ is the energy of the first excited state, ... , and $E_n$ is the energy of the $(n-1)^{th}$ excited state. (See Fig. 19.1). Here, we refer to the higher energy orbits of the electrons of the atom as excited states. These electrons have absorbed energy, but do not retain the additional energy very long. They shortly lose the excess energy by radiating one or more photons as they drop back to the ground state. Bohr showed that the energy of the nth level, $E_n$, is equal to $-hR_y/n^2$.

Bohr's scheme correctly explained the spectroscopic rules of Balmer and Ritz. Ritz's combination principle follows from the existence of atomic energy levels, the conservation of energy and Bohr's frequency condition. The model was severely limited, however, in the number of predictions it could make. For instance, there was no means of calculating the relative intensity of various spectral lines, which experimentally differed from each other. Bohr's theory was also unable to predict the polarization of the light radiated by the atom. (The polarization indicates in which direction the oscillating electric field of the photon is aligned). Finally, not all the spectral lines indicated by the model actually occur experimentally.

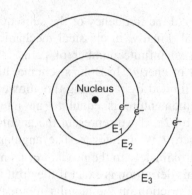

Fig. 19.1 The Bohr Atom

Bohr's model was unable to predict which transitions were forbidden. Ironically, the classical theory, which is unable to explain the observed frequencies of the atom, is, however, unlike the Bohr theory, able to calculate the relative intensity of lines, the polarization and the forbidden transitions. In order to compensate for the deficiency of his model, Bohr incorporated the positive features of the classical theory into his scheme through the correspondence principle.

Bohr noticed that for the atomic transition among the more highly excited states that the difference in energy between the two adjacent levels becomes progressively smaller as the energy increases. He also noticed that the difference in the orbital radii also becomes smaller. As one goes to higher energies the transitions from adjacent levels becomes continuous as is the case in the classical theory. This is easily seen by examining the formula for the energy, $E_n$, of the nth atomic level $En = - hR_y/n^2 = - E_1/n^2$.

The energy for all the atomic levels is negative. The reason for this is that the potential energy is negative and greater in magnitude than the kinetic energy $(1/2mv^2)$, which, naturally is positive. As long as the electron is orbiting the atom and hence, bound to it, its total energy will be negative. If its total energy ever becomes positive then it will no longer be bound to its nucleus. When referring to high-energy atomic level we will be discussing those levels for which n is large and, consequently, the energy $E_n$ is almost zero but still negative. For these high energy levels, the difference between the nth level and the (n − 1)th level is given approximately by $E_n - E_{n-1} = 2hR_y/n^3$, which goes to zero as n increases more rapidly than the energy $E_n$ itself.

Bohr also noticed that as n increases, the frequency of the photon associated with the transition from the level $E_n$ to $E_{n-1}$ approaches the frequency of the periodic motion of the level $E_n$, as expected in the classical theory. Bohr concluded, therefore, that in the limit of n approaching infinity classical physics is also able to provide a proper description of the atom. He concluded, consequently, that for large n, the quantum theory passes over into the classical theory. This statement forms the basis of Bohr's correspondence principle. He argued that if the classical theory correctly describes the frequency of the radiation for large n then the predictions of the classical theory regarding relative intensity of spectral lines, polarization and the existence of forbidden transitions would also be valid for small n.

Bohr carried the correspondence principle even further, however. He proposed that the predictions of the classical theory regarding intensity, polarization and forbidden transitions are valid for all energies. There is no theoretical justification for this extrapolation since we know that the extrapolation of the classical theory to lower energies of the radiated frequencies is incorrect. Nevertheless, Bohr's conjecture is somewhat justified on empirical grounds. It provides a fairly accurate description of polarizations. Some of its predictions regarding the relative intensities of spectral lines have also been correct but its success in this area is definitely limited.

Bohr's theory of the atom was a hodge-podge of ideas. It incorporated the concept of the quantization of energy and violated a number of the basic rules of classical theory. Through the correspondence principle, however, it still included the classical theory upon which it depended for its theory of polarization, the relative intensity of spectral lines and forbidden transitions. Despite its hodge podge nature, the Bohr model of the atom explained a surprisingly large number of the features of the spectroscopic data. Perhaps the most important prediction of the theory was the existence of discreet atomic energy levels. This aspect of the theory was dramatically confirmed one year after its formulation by Franck and Hertz.

Franck and Hertz studied the collisions of free electrons with the atoms of a gas, initially in an unexcited state. A beam of electrons with a fixed kinetic energy was directed at the gas atoms. The kinetic energy of the electrons after the collision was measured. As a result of the collision the free electrons transferred energy to the atom.

The transfer of energy from the electron can take place either by exciting the atom through an inelastic collision or by elastic scattering in which the atom remains in its ground state. In an elastic collision the total kinetic energy is conserved and in an inelastic collision it is not. Let us first consider the elastic case in which the atom is not excited. The atom will not gain very much velocity as a result of the elastic collision because its mass is so much greater than that of the electron. As a result, only an exceedingly small amount of kinetic energy is transferred from the electron to the atom and hence, the electron's energy loss is practically negligible.

The energy loss the electron suffers in the inelastic collision, on the other hand, is considerable. In this case, the atom is excited from its ground state to one of its excited levels. Because the energy levels of the atom are discreet, that is discontinuous, there is a minimum amount of energy, which must be transferred to the atom before it can be raised to one of its excited levels. This minimum amount of energy is exactly equal to the energy difference between the atom's ground state and its first excited state, $E_2 - E_1$. If the kinetic energy of the incoming electron is less than this amount, then there is no possibility of an inelastic collision since the electron does not have enough energy to raise the atom to one of its excited states. This is exactly what Franck and Hertz observed. As long as the kinetic energy of the electron was less than the threshold for inelastic collision, $E_2 - E_1$, they only observed elastic collision in accordance with the predictions of Bohr's theory. As soon as this threshold was exceeded, inelastic collisions were observed in which the electron's energy loss was equal exactly to $E_2 - E_1$. These collisions obviously corresponded to the excitation of the atom from its ground state to its first excited state.

As the kinetic energy of the electron is increased even further beyond the threshold, $E_2 - E_1$, other inelastic collisions are observed in which the atom is excited to even higher excited states. From the observed energy losses of the electrons, Franck and Hertz were able to determine the energy differences of the various atomic levels. These values were compared with the values obtained from spectroscopic data. The two sets of values for the energy levels were in complete agreement with each other. Franck and Hertz had verified Bohr's theory of the atom mechanically.

Another experiment, which demonstrated the existence of Bohr's energy levels, was devised by Maurice de Broglie (brother of Louis de Broglie, the theoretician, whose work will be discussed in the following chapter). M. de Broglie bombarded the atom with x-rays of a known energy. He then observed the kinetic energy of those electrons ejected from atom as a result of absorbing the x-rays. From the difference of the photon's energy and the electron's kinetic energy M. de Broglie was able to determine the energy levels of the atom, which were also in agreement with those obtained from Bohr's theory.

In 1913, Moseley, in England, investigated the production of x-rays. His work revealed that the charge of the nucleus increased from one element to another by one unit of charge +e. The relation of the energy of the x-rays emitted by an atom and the charge of its nucleus was found to be exactly that predicted by Bohr's theory of the atom.

The experiments of Moseley, M. de Broglie, Franck and Hertz established, beyond a shadow of a doubt, the validity of the basic concepts of Bohr's model of the atom such as the existence of energy levels and the Bohr frequency condition. As more and more experimental information was gathered, however, it became evident that Bohr's theory was not sophisticated enough to explain all the data. When the spectroscopists looked closely at the spectral lines of Balmer, they discovered that each line was actually split into a number of finer lines. This fine structure of the spectral lines was explained by Sommerfeld making use of Einstein's relativity theory. The existence of the fine structure of the spectral lines revealed that, for each of the atomic orbits of a given radius postulated by Bohr, there are actually several orbits each with the same radius and almost the same energy but different ellipsities or different values of angular momentum. The slight energy differences of these orbits arise from relativistic effects and accounts for the fine structure of each line.

For a given value of n, which determines the radius of the orbit or half the distance of the major axis of the ellipse, the possible values of the angular momentum in units of $h/2\pi$ are $\ell = 1, 2, 3 \ldots n - 1$. Studies of the splitting of spectral lines of atoms in magnetic fields revealed that the component of the angular momentum or the plane of the electron's orbit can take on $2\ell + 1$ different orientations with respect to some external magnetic field. These studies also revealed that the electron has in addition to its angular momentum an intrinsic spin of 1/2 of $h/2\pi$, which can be oriented either up or down with respect to the external magnetic field. An electron in an atom can therefore be defined by four quantum numbers, namely, n that determines the radius of its orbit, $\ell$ that determines its angular momentum or the eccentricity of its elliptical orbit, m that determines the orientation of its orbital plane and $m_s$ that determines the orientation of its spin. The number n is an integer, which does not exceed 7; $l$ is an integer ranging from 0 to n − 1; m is an integer with $2l + 1$ values that range from $-l$ to $+l$ and $m_s = \pm 1/2$.

All electrons are identical. There is no way of distinguishing one electron from another. If an electron from an atom was removed and replaced by another electron with the same quantum numbers the atom would be identical. The principle of indistinguishability applies to all the other elementary particles in addition to the electron. There is no way of distinguishing one proton from another or one neutron from another. Two photons with the same frequency are identical.

In 1925, Pauli discovered on the basis of his study of the energy levels of various atoms that no atom could contain two electrons with the same quantum numbers. Thus, for example, two electrons in an atom with the same values of n, $l$ and m would be obliged to have their spins aligned in opposite directions, i.e., $m_s = +1/2$ and $-1/2$. It would be impossible to introduce into this atom a third electron with the same value of n, $l$, and m as the first two. The electron would be excluded by the Pauli exclusion principle. The Pauli exclusion principle also explains the regularities of the periodic table of chemical elements.

In 1871, the Russian chemist Mendeleeff proposed a classification scheme of the chemical elements. He found that if he ordered the chemical elements by their atomic mass or weight the elements with similar chemical properties recurred periodically at more or less regular intervals. He constructed a table of the elements by increasing the atomic weight such that elements with similar chemical properties appeared in the same column. He left certain entries blank, which he predicted would be filled by elements that had not yet been discovered. He also described the chemical properties these missing chemical elements would possess on the basis of the chemical properties of the other elements in its column. The discovery of these elements displaying the properties predicted by Mendeleeff dramatically demonstrated the validity of his scheme. The x-ray work of Mosely in which he determined the nuclear charge of each element helped to refine the Mendeleeff classification scheme. For example, it explained why Argon, with atomic weight 39.9 but atomic number 18 comes before potassium with atomic weight 39.1 but atomic number 19.

The chemical properties of an atom are determined by the behaviour of its electrons, principally the outer electron, which interact electro-magnetically with the outer electrons of other atoms to form chemical bonds. The electrons in the atom arrange themselves in shells. Each shell corresponds to a different value of the principle quantum number, n. All the electrons in a given shell have the same average radius, $R_n$. the electrons in an atom under normal conditions settle into the state of lowest energy, the ground state. The inner shells are therefore the first to be filled. If it were not for the Pauli exclusion principle, all of the electrons of an atom would be in the n = 1 shell. Because of the Pauli exclusion principle, however, only a certain number of electrons are allowed in each shell. First of all, only those states with orbital angular momentum equal to $\ell = 0$, 1, ... , n – 1 are allowed in the n shell. For each value of the orbital angular momentum, $\ell$, there are $2\ell + 1$ possible values of m corresponding to the $2\ell + 1$ possible orientations of the plane of the electron's orbit. Since the quantum number $m_s$ can take on either the value +1/2 or –1/2 corresponding to spin up and down, the Pauli exclusion principle allows $2 \times (2\ell + 1)$ possible states with the same value of n and $\ell$. The n = 1 shell with only $\ell = 0$ states can therefore accommodate 2 electrons whereas the n = 2 shell with $\ell = 0$ and $\ell = 1$ states can accommodate 2 + 6 or 8 electrons. The n = 3 shell, on the other hand, with $\ell = 0$, $\ell = 1$ and $\ell = 2$ states can accommodate 2 + 6 + 10 or 18 electrons.

The first entry in the periodic table is the hydrogen atom with one electron with n = 1, $\ell = 0$. The next entry, helium, has two electrons with n = 1, $\ell = 0$ and opposite spins. The n = 1 shell is completed with helium since this shell can only accommodate 2 electrons. This explains why helium is an inert gas. The chemical properties of the atom are determined by the electrons of the outer most shell. Since a closed shell of electrons are extremely stable, those atoms whose outermost electrons form a closed shell are chemically inactive and correspond to the inert gases such as helium, neon, argon, etc. The hydrogen atom with only one electron is, therefore, chemically quite active.

The third entry in the periodic table is lithium with 2 electrons in a closed n = 1 shell and a third electron in the n = 2 shell with $\ell = 0$. Lithium, an alkali metal, is extremely active chemically as are all the other atoms, which consist of closed shells of electrons plus one extra electron. These entries in the periodic table are also chemically extremely active alkali metals such as sodium, potassium and cesium.

The fourth entry in the periodic table is beryllium consisting of a closed $n = 1$ shell and two $n = 2$, $\ell = 0$ electrons. Beryllium, an alkaline earth, is quite active chemically entering into chemical bonds in which it can surrender its two outermost electrons. The next 6 entries in the periodic table, boron, carbon, nitrogen, oxygen, fluorine and neon correspond to the atomic states created by adding one more additional $n = 2$, $\ell = 1$ electron to the preceding entry. Since the exclusion principle allows only 6 $n = 2$, $\ell = 1$ states by the time we arrive at neon, the $n = 2$ shell is closed and naturally, we discover that neon is an inert gas. Fluorine, which is missing only one electron to form a closed shell, is, as expected, extremely active chemically. The same is true of the other halogens such as chlorine, bromine and iodine, which, like fluorine, are each one electron short of a closed shell.

The valency of an element is the number of electrons it gives up (if the valency is positive) or gains (if the valency is negative) when it enters into chemical combinations. Those elements with positive valencies, which tend to readily give up electrons easily in chemical reaction, also tend to be good electrical conductors. The valence number corresponds exactly to the number of electrons in excess of or missing from a closed shell as a study of the last eight entries to the periodic table reveals. Lithium, beryllium, and boron have the valencies +1, +2 and +3, respectively, as well as 1, 2 and 3 extra electrons in addition to their closed shell of two $n = 1$ electrons. Nitrogen, oxygen, and fluorine have the valencies −3, −2 and −1 respectively, and are missing 3, 2 and 1 electrons from the $n = 2$ shell. Neon, with an $n = 2$ closed shell, has a valency of zero. This leaves carbon, which is more complicated since carbon can be viewed as having four electrons in excess of its closed $n = 1$ shell or missing four electrons from its $n = 2$ shell. We would expect on the basis of this, therefore, that carbon would enter into complicated bonds involving valencies of either +4 or −4. This is indeed the case, which explains the extremely complicated chemistry of the carbon atom and why it is able to form molecules with long chains.

The next eight entries of the periodic table after neon correspond to the filling of the $n = 3$ shell. First, the two $\ell = 0$ states are filled and then the 6 $\ell = 1$ states. The eight outer electrons of the $n = 3$ shell of argon, an inert gas, form a closed shell. The eight elements, sodium, magnesium, aluminum, silicon, phosphorous, sulphur, chlorine and argon follow exactly the same pattern of chemical properties as the previous eight entries, with the valencies +1, +2, +3, ±4, −3, −2, −1 and 0, respectively.

The next entry in the periodic table does not continue to fill the n = 3 shell by populating the $l = 2$ states as might be expected. Instead, the two n = 4, $l = 0$ are populated in the next two entries, corresponding to potassium and calcium. These two elements have the valencies +1, and +2, respectively, since they have 1 and 2 electrons in excess of their closed n = 3 ($l = 0$ and $l = 1$) shell.

The n = 4, $l = 0$ states were filled before the n = 3, $l = 2$ states because they are lower energy states in spite of the fact that their average radius is larger than the n = 3, $l = 2$ states. Because the orbit of the n = 4, $l = 0$ states are much more elliptical than the n = 3, $l = 2$ whose orbit is essentially circular, the n = 4, $l = 0$ electron penetrates the inner shells of electrons and hence, feels the influence of the positive charge of the nucleus more strongly. The electrons of the n = 3, $l = 2$ circular orbit, on the other hand, are almost completely shielded from the positive charge of the nucleus by the electrons of the inner shell. The magnitude of potential energy of the n = 4, $l = 0$ electrons is greater, therefore, than that of the n = 3, $l = 2$ electron. Since the potential energy of the attractive force is negative, the total energy of the n = 4, $l = 0$ electrons is less than that of the n = 3, $l = 2$ electron. The dependence of the energy of the electron on the ellipsity of the orbit and hence, angular momentum described above also explains why the $l = 0$ states are always filled before the $l = 1$ states within the same n shell.

Once the two n = 4, $l = 0$ orbits have been filled, the n = 3, $l = 2$ orbits are then populated, which accounts for the next ten entries of the periodic table, scandium, titanium, vanadium, chromium, manganese, iron, cobalt, nickel, copper and zinc. Each of these elements are metals that are sharing many chemical properties, which are mainly determined by their two n = 4, $l = 0$ electrons. The next 6 entries populate the n = 4, $l = 1$ orbits producing another closed shell with krypton and completes period 4 of the periodic table. Period 5 of the table repeats the pattern of period 4, which the successive population of the two n = 5, $l = 0$ states, the ten n = 4, $l = 2$ states and finally the six n = 5, $l = 1$ states to form another closed shell. Period 6 of the table is more complicated consisting of 32 entries instead of the 18 entries of the two preceding periods. The extra entries are due to the inclusion of the fourteen n = 4, $l = 3$ states corresponding to the rare earth metals. The following states are successively populated in period 6: the two n = 6, $l = 0$ states, the fourteen n = 4, $l = 3$ states, the ten n = 5, $l = 2$ states and finally, completing a closed shell the six $n = 6$, $l = 1$ states. Period 7 begins to repeat the pattern of period 6 but never completes it because there are not enough elements. To date, 105 elements have been observed, 92 of which occur in nature and 13 of which, the transuranic elements, have been created artificially in the laboratory. The following states are successively filled in period 7: the two $n = 7$, $l = 0$ states, the fourteen $n = 5$, $l = 3$ states and finally, three of the $n = 6$, $l = 2$ states, which accounts for the presently, observed elements.

The Bohr theory of the atom explained a great deal regarding the spectrographic and chemical properties of the atom. It had serious shortcomings, however. In spite of the correspondence principle it was never really able to successfully calculate the probability for a particular atomic transition and hence, could not account for the relative intensity of spectral lines. There were also other aspects of the spectroscopic data that could not be explained within the framework of Bohr's model. Although Bohr's model was to be replaced with the more sophisticated wave mechanics, many features of Bohr's theory would be retained in the new quantum theory. Bohr's theory had been invaluable; for many years it provided a framework for organizing the wealth of information gathered by the spectroscopists.

Before passing on to wave mechanics I would like to share with the readers my 15 minutes of fame when I had lunch with Niels Bohr in 1957. I was a first year student at MIT and Bohr was on campus to give some lectures. I saw him enter the cafeteria by himself. I jumped up from my table and ran to the cafeteria line and stood exactly behind him. Timidly I said to him, "Good afternoon Prof. Bohr." He wheeled around and said to me, "How do you do? I am Niels Bohr." I said hello and said, "I am Bob Logan". He asked me if I would like to join him for lunch, which I did. I sat there for those 15 minutes conversing with the great man himself. He was very kind and inspiring exactly as he was known by his reputation. Every time I would teach the Bohr atom as a professor in later years I would always tell this story to the delight and inspiration of my students. In subsequent years I met his son Aage Bohr at Los Alamos and only just recently in August of 2009 his grandson Tomas Bohr in Copenhagen, all physicists and just as kind and inspiring as their famous father and grandfather.

# Chapter 20

# Wave Mechanics

If quantum mechanics hasn't profoundly shocked you, you haven't understood it yet. – Niels Bohr

In the early 1920's while most atomic physicists were concerning themselves with different aspects of Bohr's model of the atom, Louis de Broglie, working essentially in isolation on his doctoral thesis at the Universite de Paris, broke new ground. Louis de Broglie was concerned with the question of the wave-particle duality of light. Light had classically displayed the behaviour of a wave, which the diffraction and interference phenomena studied by Fresnel and Young had revealed. Einstein's description of the photoelectric effect, and the subsequent discovery of the Compton effect, however, had also revealed the particle nature of light. De Broglie tinkered with the notion that the photon might possess an unobservingly small mass. This perhaps led him to speculate on the possible wave-particle dual nature of elementary particles such as electrons. De Broglie dropped the notion of a photon with mass. He proposed, however, that particles like electrons might possess wave behaviour. He based his conjecture on the fact that the photon was both a wave and a particle. Why shouldn't the same be true for an electron?

To determine the frequency and wavelength of the electron wave, de Broglie borrowed directly from Planck and Einstein's concept of the photon. According to their quantum hypothesis, the energy of a photon is equal to Planck's constant, h, times its frequency, f so that $E = hf$. De Broglie assumed the identical relation held for the electron and hence, the frequency, f, of an electron is equal to its energy, E, divided by h. There still remained the assignment of the electron's wavelength. The wavelength of the photon, $\lambda$, is related simply to its momentum, p, by the formula, $\lambda = h/p$. This follows from the fact that the momentum of the photon $p = E/c = hf/c = h/\lambda$ where we used the definition $\lambda = cT = c/f$ where $T = 1/f$ is the period of the wave.

Although the relation between momentum and energy for the electron is not the same as the photon, de Broglie assumed that the relation between momentum and wavelength are the same. He, therefore, concluded that the wavelength of the electron, $\lambda$, was also equal to $h/p$. De Broglie applied his hypothesis of the wave nature of the electron to the problem of Bohr's atom and in particular to the question of how stable orbits could be formed whose angular momentum was just equal to an integer times $h/2\pi$. De Broglie assumed that the electrons orbiting the nucleus of the atoms formed standing matter waves.

Let us digress for a moment and consider the standing waves of a violin string, which is pinned down at its two ends. The violin string can only vibrate in certain modes called standing waves. The condition on the vibration is that an integer number of half-wave lengths fit into the length of the string. Only these vibrations will reinforce themselves, after they are reflected from the ends of the strings. Vibrations with different wavelengths will interfere with their reflections from the end points and quickly die out. Fig. 20.1 shows the standing waves for the three simplest modes.

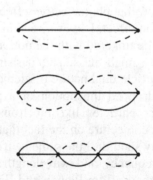

Fig. 20.1 Three Standing Waves.

De Broglie argued that only those electron waves, which formed circular standing waves and hence, could reinforce themselves would form orbits. The condition for forming a circular standing wave is that the wavelength fits into the circumference of the orbit an integer number of times. In this way, the crest or maximum of the electron wave after circling the orbit one time would match up with another crest and reinforce itself. In the same manner, the trough or minimum of the wave would also match after orbiting the circle one time. The condition that the wavelength fits an integer number of times into the circumference, c, of the orbit of radius R, is given by the formula: $c = 2\pi R = n\lambda$.

The wavelength, $\lambda$, is related to the momentum by $\lambda = h/p$. Hence, the condition for stability becomes $2\pi R = nh/p$. Rearranging the terms of this equation we have $L = pR = nh/2\pi$, since the angular momentum for a circular orbit is just p times R. This condition is exactly Bohr's angular momentum condition that we encountered in the last chapter.

De Broglie's application of his wave hypothesis to the electron enabled him to explain a key feature of Bohr's theory of the atom. His hypothesis regarding the wave nature of the electron applies to all particles. We must, therefore, ask why the wave nature of particles had not been observed before de Broglie made his hypothesis. Let us first consider the wavelength of a macroscopic object, for example, a baseball that has just been thrown by a player. The mass of the ball is approximately 0.2 kg and its velocity is approximately 300 m/sec. The wavelength of the ball is therefore only $10^{-21}$ cm.

With wavelengths as small as $10^{-21}$ cm one would never expect to detect any wave behaviour from a macroscopic object. It is only when we get to atomic size particles that we could possibly observe any wave behaviour because it is only for these small size objects that the wavelengths become large enough to detect wave behaviour. De Broglie, therefore, predicted that only subatomic particles such as electrons would display wave behaviour.

The only way to test de Broglie's hypothesis was to observe the wave behaviour of the electron. This did not happen immediately. Although de Broglie was able to explain Bohr's frequency condition, his contemporaries were quite skeptical about his result. They thought his work was a wild theoretical scheme unrelated to reality. There was even some doubt as to whether or not his work would be accepted for his doctoral thesis. His thesis was finally accepted and ironically turned out to be the only doctoral thesis, which won its author both a doctor's degree and a Nobel Prize. Luckily, de Broglie's work was brought to the attention of Einstein who was very favorably impressed by it. Einstein, in turn, passed the thesis on to others. To Max Born he remarked, "Read it even though it might look crazy, it is absolutely solid." With Einstein's blessings the experimentalists began a systematic search to detect the wave nature of the electron suggested by de Broglie.

Several physicists around the world began working on the problem. One of the groups consisting of Davisson and Germer attempted to detect the wave behaviour of particles by scattering electrons off of nickel. They were off to a flying start when their apparatus blew up and they had to start practically from scratch. Luckily they were able to salvage their target by treating it with heat. Their misfortune was actually a blessing in disguise. As a result of the heating, the nickel target had crystallized. When Davisson and Germer returned to their experiment, they immediately observed a diffraction pattern identical to the one that had been observed in earlier x-ray scattering experiments. There was no doubt about it. The electron behaved exactly like a wave. It produced the same diffraction pattern as x-rays. The paradox of quantum physics had come full circle; not only did waves display particle behaviour but now particles displayed wave behaviour. The wave behaviour of the electron was displayed in other experiments following Davisson and Germer's success. The wave behaviour of neutrons was also discovered in a similar diffraction scattering experiment with a crystal.

Long before Davisson and Germer had completed their experimental demonstration of the wave nature of the electron, an Austrian physicist named Erwin Schrödinger began developing de Broglie's ideas of standing matter waves. De Broglie's work had been brought to his attention by Einstein whom he thanks in a letter dated Zurich, April 23, 1926: "Besides, the whole thing would certainly not have originated yet, and perhaps never would have, (I mean not from me), if I had not had the importance of de Broglie's ideas really brought home to me by your second paper on gas degeneracy." Schrödinger developed de Broglie's notion of standing matter waves to a much greater extent. De Broglie's standing waves were essentially one-dimensional circular waves. Schrödinger considered the electron as a cloud, which filled the entire space around the nucleus of the atom and vibrated as a three-dimensional standing wave. Schrödinger developed a wave equation to describe the electron, which took into account the electromagnetic force exerted by the proton on the electron.

There is an amusing story told by Dirac related to Schrödinger's discovery of his famous equation. It seems that when Schrödinger was originally developing his ideas the very first equation he derived was a relativistic one. This equation has a number of complications. Consequently, when Schrödinger applied this equation to the problem of the hydrogen atom he was unable to obtain the desired experimental results. He was greatly discouraged and depressed and dropped the whole project for a number of months. When he returned to his work he realized that if he dropped the relativistic equation and ignored relativistic effects due to the electron's motion that a non-relativistic reversion of his original equation produced the desired results.

Schrödinger's equation is a non-relativistic equation. Relativistic corrections to his equation were developed later by Dirac who discovered the source of Schrödinger's original problems. Schrödinger encountered difficulties because he had attempted to make two steps at one time. This rarely happens in physics. Progress is usually made one step at a time. Schrödinger developed de Broglie's idea, who in turn, had developed Einstein's idea, who, in turn had developed Planck's idea.

De Broglie's standing wave formalism only allowed him to reproduce the Bohr frequency condition. Schrödinger's equation allowed him to calculate the exact energy of the atomic levels. Because his equation included the effects of the electromagnetic potential on the electron, he was also able to derive the corrections to the Bohr levels due to an external magnetic field (Zeeman effect) or due to an external electric field (Stark effect).

In addition to giving excellent agreement with spectral data Schrödinger's results helped to explain the nature of the electron's behaviour in the atom. Instead of the picture of the electron jumping discontinuously from one quantum state to another, a new view developed. An atomic transition from one level to another was seen as a transition from one standing wave configuration to another. The image of the standing wave also helped to explain why the orbits were quasi-stable and why only certain orbits were allowed.

As with any other new development in physics, Schrödinger's results provided the solution to a number of problems but only at the price of raising new problems. The foremost question was the interpretation of the standing waves. What were they in fact waves of? Schrödinger at first considered the electron literally as a material wave whose dimensions were given by that of the standing wave. It became apparent, however, that the standing wave represented a cloud of probability. That the actual dimensions of the electron were quite small and that only its probability of being detected was spread out through space. These probability standing waves that are spread out over space do not represent the electron spread out over space rather they represent the probability amplitude of finding an electron, which when found is just a point particle. Max Born was the first to make the probabilistic interpretation of Schrödinger's results. He was most likely influenced by the 1924 work of Bohr, Kramer and Slater in which they claimed that the electromagnetic wave represented the probability of detecting a photon. Schrödinger resisted the probabilistic interpretation at first because he felt that he had eliminated the discontinuity of quantum jumps by considering the electron as a material wave. After Bohr had finally convinced him that his theory was correct but his interpretation not valid, Schrödinger remarked in frustration, "If one has to stick to this quantum jumping, then I regret ever having gotten involved in this thing".

Six months prior to Schrödinger's developments in the summer of 1925 Heisenberg independently developed a completely different approach to atomic theory. His mathematical description of the atom known as matrix mechanics also dealt with probabilities. Heisenberg argued that atomic theory should only deal with observable i.e. quantities that can be directly measured. He, therefore, developed equations for the probabilities that an atom would make a transition form one quantum state to another. Heisenberg formulated his equation in the quantum domain for which Bohr's correspondence principle was valid. This enabled him to exploit classical physics, which is still valid in this domain. Using his equations, Heisenberg was able to calculate correctly the probabilities of transitions from one atomic level to another, as well as the energies of each level.

Schrödinger was able to show that his wave mechanics and Heisenberg's matrix mechanics were mathematically identical. Schrödinger's formulation of quantum mechanics proved to be more convenient for actual calculations. Heisenberg's contribution was just as important, however. Although calculations within Heisenberg's matrix mechanics were clumsier, his scheme proved extremely useful from a theoretical point of view. The relativistic formulation of quantum mechanics to be discussed later was developed by Dirac using the Heisenberg picture. Heisenberg's matrix mechanics like Schrödinger's wave mechanics was non-relativistic.

Heisenberg's formulation of quantum mechanics also leads naturally to the Heisenberg uncertainty principle. This principle has played a crucial role in understanding the physical ideas behind quantum mechanics and has in itself led to a number of developments in atomic physics. The uncertainty principle states that it is impossible to make an exact determination of both the momentum and position of a particle no matter how accurately they are measured. The uncertainty principle follows from Heisenberg's equations. Once formulated, it is easy to demonstrate that it arises from general considerations of measuring atomic phenomena and indeed, explains the necessity of a probabilistic description of atomic processes. We shall return to this question later, but let us continue our description of the uncertainty principle.

The uncertainty principle states that there is an intrinsic built-in theoretical limitation to how precisely one may measure both the position and the momentum of a particle. It is possible to measure either the momentum or the position of the particle as accurately as one cares to, however, one's measurement of the other variable suffers as a consequence. For instance, it one reduces the uncertainty in one's measurement of the particle's momentum one automatically increases the uncertainty in the measurement of its position. Labeling the uncertainty in the measurement of the momentum, p, and the position, x, by $\Delta p$ and $\Delta x$ respectively, the mathematical expression of the uncertainty principle takes the following form: the product of $\Delta p$ times $\Delta x$ is always greater than or equal to h, Planck's constant, i.e., $\Delta p \, \Delta x \geq h$.

It is obvious from this formulation that if $\Delta p = 0$ then $\Delta x$ becomes very large or vice versa. In fact, if one knows the momentum precisely such that $\Delta p = 0$ then $\Delta x$ becomes infinite, which means one loses all information about the position.

Heisenberg showed that the uncertainty principle also applies to the measurements of the energy, E, and the lifetime, t, of a system. The measurement of one interferes with one's knowledge of the other. If $\Delta E$ and $\Delta t$ are the respective uncertainties of the energy and time measurements then the uncertainty principle states that the product of these uncertainties will always be greater than or equal to h or $\Delta E \, \Delta t \geq h$.

Many physicists and lay thinkers found the uncertainty principle a complete anathema. They could not conceive how theoretical limitation to measurements could possibly be imposed upon physics. They were also offended by the probabilistic nature of the quantum mechanics, which the uncertainty principle seems to epitomize. The uncertainty principle, to my way of thinking, on the other hand, represents a natural limitation to the study of microscopic quantities whose energy is quantized. The uncertainty principle helps one understand why a probabilistic description of atomic processes is necessary.

In order to describe a physics system, which, after all, is the object of physics, it is first necessary to observe or know the system, i.e. to be able to gather information about the system. In order for the information to be useful, we would like to be certain that the act of making one's measurement on the system does not alter it to the extent that we are no longer dealing with the same system. Otherwise, we will collect information of successively different systems and never be able to describe the original system. For example, suppose I wish to know both the position and momentum of a body at the same time. If each measurement of the position imparts some unknown momentum to the body I will never be able to measure both its momentum and position simultaneously. For the measurements of macroscopic bodies this has never been a problem. One could always arrange to measure the position of a large body without affecting its momentum.

Let us consider the determination of the position and momentum of an automobile, for example. If I were to determine the position of an automobile by crashing another automobile into it then I would certainly change the original auto's momentum in some undetermined manner making the precise measurement of its original momentum impossible. But I do not have to make my measurement in such a heavy-handed manner. For instance, I could throw a tennis ball at the car and the change in the momentum I would produce would be almost completely negligible. If I wish to be even more discreet about my measurement I can make my measurement of the car's position visually. Even in this case, I will affect the car momentum ever so slightly since a visual measurement involves bouncing light off the car into my eyes.

As we know, light carries momentum so even in this case we impart some unknown momentum to the car. This effect is completely negligible when one takes into account that the momentum of the car is $10^{30}$ times the momentum of a photon of visible light. This can also be seen by examining the uncertainty principle mathematically for the case of an automobile whose length, mass and velocity are approximately 3 meters, 1000 kg and 30 m/sec respectively. The uncertainty principle states that $\Delta p\, \Delta x = m\, \Delta v\, \Delta x = h = 6.6 \times 10^{-34}$ kgm$^2$/sec.

If we divide the uncertainty evenly between the momentum and the position, the uncertainty principle prevents us from measuring the length of the car or its velocity more accurately than one part in $10^{19}$. Since the accuracy for making measurements is much less than this and since the accuracy needed to describe a system does not have to be anywhere near one part in $10^{19}$ the uncertainty principle has absolutely no effect of the description of a macroscopic system like an automobile.

Let us now consider the limitations the uncertainty principle imposes for the case of an electron in an atom. Planck's constant will no longer be such a small number. Instead of distances, like 3 meters for the automobile, we must now consider distances of the order of $10^{-8}$ cm. Instead of a mass of 1000 kg the mass of the electron is $0.9 \times 10^{-30}$ kg, and hence, its momentum is considerably less than that of the automobile. For an electron with a velocity of 0.1c the product of its position times its momentum ($10^{-8}$ cm $\times 0.9 \times 10^{-30}$ kg $\times 0.1 \times 3 \ 10^8$ m/sec) is approximately $27 \times 10^{-34}$ kgm$^2$/sec or just 4 h. It is clear that the uncertainty principle imposes severe limitations on how accurately and the momentum and position of an electron may be determined since the uncertainty is the same order of magnitude as the quantities to be measured.

Let us consider physically what is involved in determining the position and momentum of an electron. In actuality, it is not much different than measuring the position and velocity of an automobile by crashing another automobile into it. There are no particles smaller than an electron. Therefore, if we wish to detect the electron using another particle the best we can do is to use another electron. We cannot chop an electron into a thousand pieces and use a tiny chunk of an electron as a detector. We are obliged to use another electron or a larger particle.

The only other alternative is to use a photon. This presents a problem as well because the photon also carries momentum. This problem can be minimized by choosing to use a low energy and consequently a low momentum photon. The only difficulty with a low momentum photon is the fact that it will have a large wavelength, $\lambda$, since $\lambda = h/p$.

The size of a photon with wavelength $\lambda$ is at least equal to $\lambda$ and hence, the detection of the electron's position with a photon of wavelength $\lambda$ will automatically introduce an uncertainty of at least $\Delta x = \lambda$. The uncertainty in momentum inherent in the measurement is just the momentum of the photon, hence $\Delta p = h/\lambda$. The product of the uncertainties in the position and momentum is, therefore, $\lambda$ times $h/\lambda$ or h, i.e. $\Delta p \, \Delta x = h/\lambda \times \lambda = h$.

Thus, we see in accordance with the uncertainty principle that h is the minimum value for the product of the uncertainties of the position and the momentum. There is no way of avoiding the uncertainty principle. In a discussion with my students it was suggested that the uncertainty in momentum introduced by the momentum of the photon could be avoided by shooting photons at the particle equally from all sides. This ingenious proposal would avoid the uncertainty in momentum, however, in order to direct the photons equally on all sides one would have to know the position of the particle. If we could do this we wouldn't have to make the measurement in the first place.

The uncertainty principle is not really such a mysterious concept. In simple terms, it states that in order to obtain information about a system, it is necessary to disturb the system making it impossible to ever obtain a complete set of information about the system. To know something is to interact with it and hence, change it. If we apply this concept to other areas of human study it will seem even less mysterious. Let us consider the problem of a biologist studying the behaviour of a group of animals. The biologist is well aware that his observations will be contaminated to some extent since the animals will be aware of being observed and behave differently than they would in a total state of nature. One can minimize the effects of such observations. Jane Goodall's observations of chimpanzees made while she lived in the wilds with her subjects is far more accurate than those made by observing chimpanzees in a zoo. In spite of all her care and trouble, the behaviour of the chimpanzees observed by Jane Goodall was still influenced to some extent by her presence. The concept of the uncertainty principle can also be applied to interpersonal relations. If I want to get to know a person, then I must interact with them and hence, change them to some extent.

The limitations of the physicists' knowledge of his physical world arise more or less for the same reason that they arise in the field of biology of interpersonal relations. To know something is to interact with it and hence, disturb it. Perhaps the same reason that so many people (both scientist and lay people) find the uncertainty principle so disturbing is that a myth has developed surrounding physics. As a result of the success of Newtonian physics people began to believe that the physicists were able to provide an exact mathematical description of the physical world. The success of the post-Newtonian physics in describing other physical phenomenon such as electricity, magnetism, heat, sound and light help to reinforce this myth. Thus, it was a great shock to scientists and non-scientists alike when the study of atomic physics revealed that there are limitations to man's knowledge. We shall return to some of the philosophical implications of the uncertainty principle raised here but let us first examine its implications for physics.

In classical or Newtonian physics it is possible to determine the exact position and momentum of a body. Once this information is known it is then possible to specify the exact position and momentum of the body for all future times as long as one knows the forces acting on the body. This programme is no longer possible within the framework of quantum mechanics even is one knows all the forces on the body. The uncertainty principle only permits a partial or approximate knowledge of the location and momentum of the body. For instance, one might know that the body is between x and x + Δx with a momentum between p and p + Δp where Δp Δx = h. It is now impossible to specify exactly where the particle will be some time later. A particle at x with momentum p will behave differently than a particle at x + Δx with momentum p + Δp. The uncertainty will tend to increase with time.

The uncertainty principle forces a change in our description. Instead of specifying the exact location and momentum of the particle as a function of the time as was done in classical physics, we are now forced to describe the particle in terms of the probability of finding it at some place, x, with a momentum, p. Within the framework of quantum mechanics, the probability of determining the particle's position and momentum is described quantum mechanically by the probability amplitude or wave function, $\psi(x,y,z)$, which has a unique value for each position in space x, y and z.

The equations describing the behaviour of the wave function, $\psi$, is the Schrödinger equation referred to earlier. When Schrödinger first discovered his equations he did not connect the wave function, $\psi(x, y, z)$ with probability but rather interpreted it as the density of the electron cloud, which he considered to be spread out through space. It was Max Born who pointed out that the correct interpretation was to continue to assume that the electron is a point particle and to regard $\psi$ as a measure of the probability of detecting the point electron at some point in space. He showed, in fact, that the probability of finding the electron located at the point in space x, y and z is just the absolute value of the wave function multiplied by itself, $|\psi(x, y, z)|^2$. He also showed that Schrödinger's $\psi$ determines the probability that the particle has a particular value of the momentum can also be determined from $\psi$ but involves a more complicated mathematical operation than multiplying $\psi$ by itself. These mathematical details need not concern us. The important point is that once one knows a particle's wave function, $\psi(x, y, z)$, one can determine the probability it will have a particular momentum and a particular position.

The wave behaviour of the electron is due to the fact that its probability amplitude, $\psi$, behaves like a wave. The Schrödinger equation is basically a wave equation. The diffraction pattern observed when a beam of electrons passes through two slits is due to the interference of the wave function passing through the two slits with itself. It is not due to the interference of a beam of electrons passing through one slit with the beam of electrons passing through the other slit. This has been experimentally demonstrated by reducing the flux of electrons so that only one electron at a time passed through the two-slit system. The position on the screen where the electron landed after passing through the slits is recorded and tabulated. After a sufficient time passes the pattern that emerged was the same characteristic diffraction pattern one obtained with a high flux beam of electrons. There were positions on the screen where electrons would go only if both slits were open and would not go it slit 1 was open and slit 2 was closed or if slit 1 was closed and slit 2 was open. There were also positions where the electrons would not go if both slits were open but would go if one or the other slit was open. This is extremely mysterious. In the latter case I open slit 1 and close slit 2 and observe electrons at position Y on the screen. If I now open both slit 1 and slit 2, electrons no longer go to position Y. It is impossible to understand in terms of a particle how the opening of slit 2 suddenly prevents electrons from going to position Y via slit 1.

In view of the fact the results hold even with one electron at a time the only way of interpreting this is to assume that the electron passes through both slits and interferes with itself. But how is this possible if the dimensions of the electron are smaller than the distance between the two slits? The way this phenomenon is understood in terms of our quantum mechanical description is to recognize that the probability amplitude, $\psi$, describing the electron is interfering with itself.

Let us call the wave function at slits 1 and 2, $\psi_1$ and $\psi_2$, respectively. If only slit 1 is open then one obtains the pattern given by $|\psi_1|^2$ and if slit 2 is only open one obtains the distribution given by $|\psi_2|^2$ as is shown in Fig. 20.2(a). If both slits are open then the probability of finding the electron is given by $|(\psi_1 + \psi_2)|^2 = |\psi_1|^2 + \psi_1^* \psi_2 + \psi_1 \psi_2^* + |\psi_2|^2$.

This distribution is shown in Fig. 20.2(b), and is not simply the sum of patterns due to slit 1 and slit 2 being open separately. The reason it is not simply the sum is because of the interference of the wave functions at slit 1 and 2, namely, $\psi_1$ and $\psi_2$. The square of the sum $\psi_1 + \psi_2$ contains more than just the terms $|\psi_1|^2$ and $|\psi_2|^2$ but also the interference terms $\psi_1^* \psi_2 + \psi_1 \psi_2^*$, which are responsible for the diffraction pattern. It is the presence of these terms that explain how it is possible that certain positions on the screen are struck by electrons with only one slit open but receive no electrons with both slits open. Only a probabilistic description allows us to understand how electron diffraction takes place.

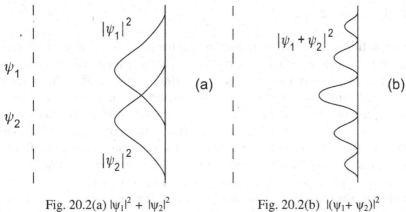

Fig. 20.2(a) $|\psi_1|^2 + |\psi_2|^2$          Fig. 20.2(b) $|(\psi_1 + \psi_2)|^2$

## Wave Packets

De Broglie was the first to indicate the wave-like properties of the electron, which explains the diffraction behaviour described above. De Broglie assigned to the electron a single wavelength $\lambda$ related to its momentum p by $\lambda = h/p$. We have learned, however, in our discussion of the uncertainty principle that a particle rarely has a precisely defined momentum. We therefore expect the wave function describing the particle to be composed of many waves, each with a different wavelength. Such a superposition of waves is called a wave packet. Let us consider a wave packet composed of several waves with wavelengths between $\lambda$ and $\lambda + \Delta\lambda$. The momentum of these waves and hence the particle is spread out between $h/\lambda$ and $h/(\lambda + \Delta\lambda)$. The uncertainty in the momentum of the particle is therefore approximately $\Delta p = h/\lambda - h/(\lambda + \Delta\lambda) = h \, \Delta\lambda/\lambda^2$.

The size of the wave packet describing an atomic particle increases with time. This does not mean that the actual size of the particle increases. It remains the same but the wave packet and hence, the uncertainty in the particles position does increase with time. The reason for this is that the different waves in the packet are traveling at different speeds because of their spread in momentum. The waves with momentum $p + \Delta p$ gets ahead of the waves with momentum p and hence, there is an increase in the size of the wave packet. This surprising result is a unique feature of quantum mechanics not to be found in classical physics.

## Barrier Penetration

Perhaps the most baffling quantum mechanical effect of all is barrier penetration or tunneling. Let us consider a particle under the influence of a spherically symmetric force with a radius $r = r_0$ so that the particle is trapped within the space $r < r_0$. The potential energy of the particle under the influence of the force is $-V_0$ which means the particle is trapped unless it has kinetic energy greater than or equal to $V_0$. The nature of the force is to create a barrier that prevents particles with total kinetic energy less than $V_0$ from leaving the region r less than $r_0$. From a classical point of view if a particle's total energy is less than $V_0$, the maximum potential energy of the force, it will remain exclusively in the region $r < r_0$. It will not be able to pass into the region $r > r_0$. The nuclear force that traps the protons and neutrons (nucleons) within a nucleus, is this kind of force. If the classical laws of physics applied, one would never expect a proton or neutron to escape the nuclear barrier. There are cases, however, of unstable nuclei where the nucleons escape their nucleus.

This effect can be explained in terms of quantum mechanics. The equations governing the behaviour of the wave function, $\psi$, demand that the $\psi$ become very small very quickly in the forbidden region. However, $\psi$ does not become identically zero in this region as it would within the framework of classical physics. Because $\psi$ is not identically zero in the forbidden region there is a very small but finite probability that the particle will enter this region. If the particle passes through this region to values of $r > r_0$ it will then be free of the nucleus because now its total energy is positive and hence, its wave function in this region no longer needs to remain small. In this way nucleons are able to tunnel through the barrier created by the nuclear force and escape the nucleus. An equivalent event on the macroscopic level would be the exit of a student from a schoolroom by walking through the walls leaving them intact.

The probability of the tunneling of a particle is very small because the wave function in the forbidden zone is so small. Given the potential energy due to a given force one can use the Schrödinger equation to calculate the probability of the penetration of this potential barrier. The calculation of the lifetime of unstable nuclei based on barrier penetration has given excellent agreement with experimental value confirming the quantum mechanical effect of tunneling. This effect is strictly quantum mechanical and would not be possible were it not for the uncertainty principle. When the particle passes through the forbidden zone, a violation of energy conservation takes place. The time during which this violation occurs, $\Delta t$, is very small so that the product of the energy conservation discrepancy times the time the particle is in this state is less than h. But according to the uncertainty principle the product of the uncertainties $\Delta E$ times $\Delta t$ is greater than h, hence the energy conservation violation can never be detected because measurements accurate enough to show this are not possible. This is one of the mysteries of quantum mechanics.

# Chapter 21

# Philosophical Implications of Quantum Mechanics

At the end of the last chapter we discussed several quantum mechanical phenomena, which are impossible to conceptualize from a classical point of view. The diffraction of electrons, the expansion of wave packets and the tunneling of potential barriers all involve the breakdown of causality. Causality is the concept, which forms the very foundation of classical physics. Even relativity did not change the notion of causality. In fact, causality plays an important role in the formulation of relativity. The breakdown of the concept of causality, intrinsic to the formulation of quantum mechanics by Heisenberg and Schrödinger, produced shock waves among physicists and philosophers, which persist to this day.

Although the formulators of quantum mechanics were forced to give up the notion of causality, they were still able to construct a theory of which possessed predictive powers. They could no longer describe the behaviour of an individual particle. They were able, however, to provide a probabilistic description of individual particles, which enabled them to predict the behaviour of statistically large ensembles of particles. Thus, they were able to preserve the most basic aspect of a scientific theory, namely its ability to make predictions, which can, in turn, be verified by experiment. In fact, quantum theory was able to explain a great deal of the behaviour of atoms and molecules. Perhaps the most important of these results was the description of the spectral and chemical properties of various atoms and molecules as well as the structure of matter such as solids and liquids.

The success of quantum physics as a descriptive theory of large numbers of atoms was universally accepted within the physics community. A schism developed within the community, however, regarding the interpretation and meaning of the theory. A number of scientists including Einstein, Schrödinger, Planck and later, de Broglie, were unable to accept the operation of chance within the framework of quantum mechanics. They believed that since quantum theory could not provide a causal description of the behaviour of individual particles that it was an incomplete theory. They conceded that quantum mechanics was a logically consistent scheme, which was able to describe experimental results accurately. They accepted that the limitations of measurements imposed by the uncertainty principle made a probabilistic description necessary. They believed, however, that the uncertainty principle was only a limitation of our knowledge and that, in actuality, the particle has both a precise position and velocity and that its behaviour is causally determined. They could not accept the notion that chance could actually enter into the behaviour of the physical world.

Einstein, who became the leader of this position, expressed the concern of his school of thought with his often quoted remark, "I do not believe in a God that plays dice." Einstein and the others believed in the existence of an underlying determinism, which actually guided the particles. They simply did not accept the idea that the behaviour of particles could be governed by chance. They wanted to know what was really happening to the particles. They believed a theory would eventually emerge in which the hidden causality would appear. They, therefore, considered quantum theory as incomplete and anticipated the appearance of a fuller theory, which would eventually replace it. Nothing has emerged to this day more than 80 years since Schrödinger first formulated wave mechanics.

In opposition to this point of view were the proponents of the Copenhagen interpretation who had formulated their ideas in Copenhagen in the late 1920's under the leadership of Niels Bohr and included Heisenberg, Dirac, Born, Pauli and Kramers. They regarded the new quantum mechanics as a complete theory. They believed that the uncertainty principle imposed a limitation on our knowledge of the world and hence, a limitation on the behaviour of the particles themselves. They made no distinction between reality and our knowledge of reality.

They considered the question asked by Einstein and his followers, "what is really happening to the particle as meaningless?" What can not be observed or measured is of no concern to physics since there can never be verification of a theory of unobservable behaviour. An expression of the view is found in the writings of Dirac who wrote the following: "The only object of theoretical physics is to calculate results that can be compared with experiment, and it is quite unnecessary that any satisfying description of the whole course of phenomena should be given." This controversy is being revisited today between the proponents and critics of string theory. The critics say that string theory is not physics because in its 30-year history it has never made a prediction that can be measured or observed.

One should not interpret Dirac's statement in a trivial way and conclude that he and his colleagues decided that since their results explain experimental results there was no need for further reflection. The members of the Copenhagen school were just as perplexed and disturbed by the new ideas. They, too, spent long hours trying hard to understand the underlying ideas behind their new theory. The following passage from Heisenberg's book entitled *Physics and Philosophy* indicates the nature of their frustration:

During the months following these discussions an intensive study of all questions concerning the interpretation of quantum theory in Copenhagen finally led to a complete and, as many physicists believed, satisfactory clarification of the situation. But it is not a solution, which one could easily accept. I remember discussions with Bohr which went through many hours till very late at night and ended almost in despair; and when, at the end of the discussion, I went alone for a walk in the neighbouring park, I repeated to myself again and again the question: Can nature possibly be as absurd as it seemed to us in those atomic experiments?

Coming to grips with the peculiarities of quantum mechanics was no easy task for Bohr and his co-workers. Einstein and his sympathizers in a sense avoided the problem by refusing to give the new quantum mechanics the status of a complete theory. They put the problem off by claiming that some day a complete theory would arise and explain all. Bohr, on the other hand, accepted the paradoxes of quantum mechanics even though they contradicted the intuition of physics that he and others had developed from their study of classical physics. Bohr recognized that although quantum mechanics had completely invalidated classical physics that it was still necessary to use the language of classical mechanics to describe the behaviour of atomic systems. This is perhaps the ultimate contradiction of quantum mechanics. It would be quite impossible to develop quantum mechanics in purely quantum mechanical terms. Thought by itself is a classical process. Our entire experience is classical. If the laws of quantum physics operated on the macroscopic level we would never have developed physics. Life itself would not be possible. imagine the difficult of crossing the street safely if the approaching cars are only 90% on the road and 10% of the time are on the sidewalk.

The world of our experience is classical, not quantum mechanical. It is fairly obvious the only language that we have at our disposal to describe the world whether it is the macro-world or the atomic world is the language of classical physics, even if that language is inadequate. It should not surprise us, therefore, that our description of the atomic world will be less than satisfactory at times. The description of elementary particles such as electrons is a perfect example. In classical physics, the concept of a wave and the concept of a particle were well-defined notions, which were mutually exclusive. A wave represented a phenomenon whose character was completely different than that of a particle. Within the framework of classical physics these two concepts were used to describe light and electrons, respectively. Light was a wave phenomenon and the electron, a particle and that was that. Within the framework of quantum physics, the behaviour of light and the electron is a great deal more complicated than that of a wave and a particle, respectively. There is, in fact, no way of describing their behaviour classically. The best one can do is to say that there are times when light behaves like a wave and there are times when it behaves like a particle. The same is true of the electron. The classical notions of the wave and the particle, according to Bohr, form complementary descriptions of the electron. They also form a complementary description of light. Although the wave and particle notions mutually exclude each other they are both necessary for a description of either light or the electron. This is the essence of Bohr's complementarity principle.

The classical notions of position and momentum are also complementary concepts. In classical physics, one can assign an exact value to both quantities whereas in quantum physics the more one learns about one variable the less one knows of the other. A quantum system cannot have a well-defined position and momentum at the same time. Yet, these two mutually exclusive quantities are needed to describe the quantum system. Position and momentum are complementary concepts in the same sense as the wave and the particle notion are complementary. There is a famous quote Bohr made regarding truth, which I believe contains the essence of the complementarity principle. He said: "(There are) two kinds of truths. To the one kind belong statements so simple and clear that the opposite assertions could not be defended. The other kind, the so-called "deep truths" are statements in which the opposites also contain deep truths."

Einstein and his followers could not accept the tenets of the complementarity principle with its dichotomous description in terms of waves and particles. Instead, they adopted another dichotomy. They believed in two levels of reality, the one which is observed and the other which "really is". This division is not unlike the one Kant makes between phenomena or perceived events and noumena, or things in themselves. Einstein's position with regard to the completeness of quantum theory is that while quantum mechanics presently describes phenomena accurately enough, it is incomplete because it cannot describe the noumena. Only a causal description could properly describe the noumena. The Copenhagen school denies the distinction between phenomena and noumena and like the British empiricist, believes one can only describe that which is perceived.

Einstein believes that there is an intrinsic order within the physical world, which exists outside of man. Bohr, on the other hand, believes that the order we have discovered in the physical world we have imposed through our mental activity. Einstein believes that the order was there and we were clever enough to uncover it. The difference in these two men's attitude partially explains their reaction to quantum theory. Einstein is disappointed because the order that quantum mechanics reveals is only partial, it is incomplete and hence, so is the theory. Bohr, however, seems quite delighted that he and his co-workers were able to go as far as they did and reveal as much order as they were able to find. For Einstein, the proverbial half glass of wine is half empty whereas for Bohr it is half full.

Einstein and Bohr debated their respective positions with each other for the rest of their lives. Even after Einstein had died Bohr, in his mind, was still trying to answer the question raised by the man who refused to believe that God plays dice. It way always a source of personal frustration to Bohr that he could not win Einstein to his side.

The friendly debate that arose between the two men was like none other before in the history of science. In previous debates, the conflict usually centered on which of two conflicting theories provided the best description of nature. The goals and aims of science were usually agreed upon. In this debate, the reverse was true. Both sides agreed that quantum mechanics provides an accurate description of the experimental data. What was at issue was the aims of physics. The agreement as to whether or not quantum mechanics is a complete theory arose from a difference in opinion on the basic goals of science.

Einstein argued that the aim of science was to provide a causal description of nature while Bohr maintained that the only aim was to describe the experimental data. There was no way they could really resolve their conflict. All the same, they argued back and forth for many years. Einstein was continually proposing thought experiments to demonstrate possible contradiction in the theory. Bohr was always able to answer him satisfactorily. Einstein persisted, but finally had to admit that the theory was self-consistent but would never concede that it was complete. Einstein's friend, the physicist Ehrenfest, one chastised him for his stubbornness, "Einstein, shame on you. You are beginning to sound like the critics of your own theories of relativity. Again and again your arguments have been refuted but instead of applying your own rule that physics must be built on measurable relationships and not on preconceived notions, you continue to invent arguments based on those same preconceptions." It is perhaps one of the great ironies of all physics that Einstein, the inventor of relativity, the greatest innovator physics had ever known with the possible exception of Newton was unable to accept quantum mechanics in spite of all his important contributions in the field. Bohr, probably better than any one else, understood the wisdom of the following remark attributed to von Helmholz, "The originator of a new concept finds, as a rule, that it is much more difficult to find out why other people do not understand him than it was to discover the new truths."

There is no true resolution to the Bohr–Einstein debate. The only way that the Einstein position could truly be vindicated is for someone to discover a complete theory which both describes the experimental data and restores causality. Numerous attempts in this direction have been made without any notable success. In the absence of such a theory, one is more or less inclined to agree with Bohr. Only time will tell.

One approach of the followers of Einstein's position is to say that there are hidden variables that once they are found would restore causality. All that I can conclude is that these variables are very well hidden. But the hunt for them continues, although I do not think any will be found. We just have to accept this is the way our universe is.

Chapter 22

# Quantum Electrodynamics

The quantum mechanics developed by Schrödinger and Heisenberg did not take into account the effects of Einstein's theory of relativity. The neglect of relativity did not limit the usefulness of the theory in any serious way, however. The velocity of the electron in an atom or molecule is never more that a few percent of the velocity of light and, therefore, any relativistic corrections that are necessary are extremely small. Non-relativistic quantum mechanics proved to be an extremely useful tool in understanding the behaviour of atoms and molecules and hence, in understanding the behaviour of macroscopic materials such as solids, liquids and gases. In spite of its many practical applications physicists were not completely satisfied with non-relativistic quantum mechanics. First, there were some small details, which the theory did not completely explain. But perhaps more importantly there were aesthetic reasons for wanting to combine relativity and quantum mechanics. Schrödinger, himself, as previously related had originally attempted to incorporate relativity into his wave equations but failed because of the complexity of the fully relativistic equations. Shortly after reporting his non-relativistic results, Klein and Gordon resurrected Schrödinger's original relativistic equation. They were unable to do very much with it except to write it down and point out the difficulties that attended its use.

Paul Dirac also attacked this problem and was able to develop a different equation, which was fully relativistic and avoided some of the difficulties of the Klein-Gordon equation. Dirac found an equation to describe the electron, which combined the results of both quantum mechanics and relativity. In addition, Dirac's equation also explained the spin of the electron and the presence of its magnetic properties.

Before Dirac's theory, it was impossible to explain the magnitude of the electron's magnetic momentum in terms of its spin. Any mechanical model of the electron, which attributed its magnetic moment to the rotation of the electron's charge always involved velocities greater than the velocity of light and hence, contradicted relativity. Dirac's equation, however, successfully related the magnitude of the magnetic moment of a charged particle to its spin, its charge and its mass. In addition to providing explanations for the spin and magnetic momentum of the electron, Dirac's equation was also able to explain the minute splitting of certain spectral lines, which the non-relativistic quantum mechanics had not been able to explain.

Dirac's equation was a great success. It contained one difficulty, however, the resolution of which would prove to be interesting. Dirac's equation contained two types of solutions. One set of solutions corresponded to normal electrons with positive energies greater than or equal to $m_ec^2$, the rest mass energy of the electron. Another set of solutions corresponded to electrons with negative energies less than or equal to $-m_ec^2$. The existence of the negative energy states seems quite unfeasible experimentally. If such states existed, electrons would spontaneously make quantum jumps from the positive energy states to the negative energy states emitting photons as they made this transition. Dirac excluded this possibility by a very imaginative and speculative application of the Pauli exclusion principle. He postulated that all of the negative energy electron states were occupied and that there existed an infinite sea of negative energy electrons. Transitions from positive energy states to negative energy states could not take place because the Pauli exclusion principle does not allow an electron to make a transition into a state, which is already occupied. This resolved the problem of the transition to negative energy states but then we are left with the problem of an infinite sea of negative energy electrons.

Dirac concluded that the sea of negative energy states is unobservable and corresponds to the vacuum state. The vacuum state is the state of no observable electrons. Although Dirac believed that the negative energy sea was not observable he assumed that one could observe the absence of an electron in one of the negative energy states. A hole in the negative energy sea should behave like a positively charged particle since negatively charged particles would be attracted towards the hole to fill it and hence, would seem to be attracted by a positive charge. An analogous situation is observed in the behaviour of the Chlorine atom, which is missing one electron to fill its outer shell. The missing electron in the Chlorine's shell acts like a positive charge since this gap attracts electrons to it and explains the intense chemical activity of Chlorine.

Dirac initially identified the holes in the negative energy sea with the proton. He thought that he had shown the mass of the hole was approximately 2000 times the mass of the electron. Pauli quickly pointed out that if the proton were a hole in the negative energy sea of electrons that the electron in the hydrogen atom would quickly jump into it and annihilate the hydrogen atom leaving only two photons behind to carry away the energy. The mystery of the identification of he hole in the negative energy sea of electrons was resolved experimentally a short time later by the discovery by Anderson of particles which had the same mass as the electron but opposite charge. The new particles, called positrons, or antielectrons, were detected in an experiment in which cosmic rays were observed by a cloud chamber.

Cosmic rays are high-energy particles, produced in outer space, which enter the earth's atmosphere. The cloud chamber is a device used to detect the charged particles, which leave tracks of ionization as they pass through the vapor of the chamber. By immersing the entire cloud chamber in a magnetic field and measuring the amount and direction of the curvature of the charged particle's tracks, one can determine the mass and the charge of the particle from one's knowledge of the magnetic force. Anderson observed tracks in his cloud chamber with equal but opposite curvature of the electron. These tracks could only be made by a particle the same mass of the electron but with the opposite charge. The positron was quickly identified with a hole in Dirac's negative energy sea of electrons.

If Dirac's theory about the negative energy sea of electrons was correct, then one would expect that the electrons should jump into the holes in the negative energy sea and emit radiation. In this process, both the electron and the positron would be annihilated since the electron disappears by jumping into the hole and the positron or the hole disappears by being filled. One would, therefore, expect to observe the reaction: $e^+ + e^- \rightarrow \gamma + \gamma$ where $e^+$, $e^-$ and $\gamma$ represent the positron, electron and photon, respectively. This is precisely what is observed to happen experimentally.

Once a positron encounters an electron the two particles annihilate each other leaving two quantas of light energy. The electron has jumped into the empty negative energy state releasing the energy gained by this transition as light energy. The reason two photons are created is so that momentum will be conserved. This explains why the two photons go off in opposite directions. The amount of energy created by the annihilation is just equal to the total kinetic energy of the electron-positron pair plus the rest mass energy of the pair, which equals $2m_ec^2$. Electron-positron pair annihilation not only confirmed Dirac's theory but also demonstrated the validity of Einstein's relation between mass and energy, namely, $E = mc^2$.

Dirac's theory of the electron also explains the creation of an electron-positron pair from light energy. When high-energy photons pass close to a nucleus they are frequently converted into electron-positron pairs. In the cloud chamber this observation appears as the spontaneous appearance of a pair of tracks with equal and opposite curvature. The photon does not leave a track because it carries no charge. From the initial direction of the electron and positron tracks, however, it is possible to determine the direction from which the photon came, and hence, the source which produced the photon. The energy of the photon, which creates the electron-positron pair, must be at least $2m_ec^2$ in order to provide enough energy to produce the rest mass for the two particles. The presence of the nucleus is only necessary to conserve momentum and energy.

The process of pair creation can be envisaged using Dirac's image of the negative energy sea. One may consider the photon as providing energy to an electron in one of the negative energy states and lifting it to a positive energy state. Since the electron in the negative energy sea is unobservable, it suddenly becomes visible when it is elevated to a positive energy state. The hole in the negative energy sea of electrons created by the absence of this electron also becomes visible as a positron. Pair creation is just the opposite process of annihilation. In pair annihilation an electron falls from positive energy to negative energy radiating light whereas in pair creation an electron absorbs a photon and jumps from a state of negative energy to one of positive energy.

The key concept Dirac used in interpreting his equation was the idea of the unobservable negative energy sea of electrons. Although this notion was essential in Dirac's development of his ideas, it is not absolutely necessary to the formulation of Dirac's theory. Having once gained an insight into how the electron behaves it is now possible to discard the concept of the negative energy sea and to view the Dirac equation as describing a charged spin one half particle and its oppositely charged spin one half antiparticle for which the processes of annihilation and creation occur. The solutions to the Dirac equation, which seem to correspond to negative energy can now be viewed as positive energy solutions of the antielectron. The concept of the negative energy sea of electrons can be retained, however, to aid the mind's eye in conceptualizing pair creation and annihilation. If the idea of the negative energy sea were absolutely necessary to the formulation of Dirac's ideas, we would then have to deal with the mystery of how an infinite sea of negative energy electrons can remain unobservable. The sea of negative energy electrons does not exist in real space, it is a concept in the minds of those physicists who use Dirac's equation. Hopefully, it has also become a concept in your mind as well.

In our description of the annihilation of an electron-positron pair we failed to mention the very curious intermediate state that frequency occurs just prior to the annihilation process. The electron and positron on first encountering each other interact through their electric charges and often form a bound state analogous to the hydrogen atom before annihilating. The positron plays the role of the proton in the formation of this peculiar atom, which is referred to by the name of positronium. All of the energy levels of positronium correspond to those of the hydrogen atom. All of the levels are shifted naturally because of the difference in mass between the proton and the positron. The entire Balmer series of frequencies for positronium have been observed and measured.

Once the positronium is formed, it makes radiative transitions from level to level until it reaches its ground state. It does not remain in this ground state for very long. Because of the close proximity of the electron and the positron in the ground state, they eventually make contact and annihilate each other. The annihilation of the pair can take place from any of the levels of positronium. The probability is greatest, however, in the ground state because there is the greatest overlap of the electron and positron wave functions in this state. The probability for annihilation is just equal to the overlap of the two wave functions, i.e. the probability of finding the electron and the positron at the same place.

Dirac's relativistic equation not only describes the electron but also all spin 1/2 particles and hence, describes the proton and the neutron. We shall learn more about these particles when we study nuclear physics. For the moment, however, we shall introduce them into our discussion since Dirac's theory predicted that these particles would also have antiparticles, namely the antiproton and the antineutron. The creation of a proton-antiproton pair requires approximately 200 times the energy (the ratio of the proton mass to the electron mass) needed for the creation of an electron-positron pair. The observation of the antiproton did not take place immediately following Dirac's prediction because it was impossible to create photons with the necessary energy, $2m_pc^2$, where $m_p$ is the mass of the proton.

The discovery of the antiproton had to await the development of particle accelerators, which could yield such high energies. Such a machine was built in the early fifties at the University of California at Berkeley. Shortly after the completion of this machine the long awaited antiprotons appeared exactly as Dirac's theory had predicted. These particles have the same mass and spin as the proton but opposite charge. antiprotons and protons annihilate each other to produce photons in the same way as electrons and positrons. The reaction of the proton and antiproton is more complicated, however, because of the nuclear force but the essential features of Dirac's theory still persist. Shortly after the discovery of the antiproton the antineutron was also detected. This particle has the same mass and spin as the neutron. Its charge is also equal to zero since the opposite of zero is still zero, i.e. $(-0 = 0)$. Neutron antineutron pairs are also created and annihilated just as expected by Dirac's theory.

One of the interesting features of the antiparticles is that their interactions among themselves are exactly identical to the interactions of the particles among themselves. The antiproton antiproton interaction is the same as the proton-proton interaction. A positron and an antiproton interact with each other just like the proton and an electron and hence, form an antihydrogen atom, which behaves exactly like the hydrogen atom. The antihydrogen atom has been observed and found to emit the same spectral lines as the hydrogen atom. There is no way of distinguishing matter from antimatter. Given enough positrons, antiprotons and antineutrons, one could build a world identical to our own. In fact, it is purely convention to call the material of which we are made matter and those particles, which annihilate our particles antimatter. For all we know, we are made of antimatter and the other particles are matter. There is no way of determining whether the other galaxies or clusters of galaxies in our universe are made of matter or antimatter. There are, in fact, cosmological models, which predict that half the galaxies in the universe are made of antimatter and the other half of matter.

In addition to the relativistic equation Dirac developed for the electron, he also developed an equation to describe the photon and its interaction with electrons and other charged particles. Dirac's equations gave results in excellent agreement with experiment. They described all of the basic interactions of electrons and photons and included the processes of pair creation, pair annihilation, the emission of radiation and the scattering of electrons and light. As experimental measurements of energy levels of atoms improved, it was discovered that there were extremely small discrepancies between Dirac's theory and experiment. Through the work of Tomonaga, Schwinger and Feynman in 1948, it was discovered that these discrepancies could be accounted for by calculating corrections to Dirac's theory.

The quantum electrodynamics developed by these physicists and their co-workers yielded results of unprecedented accuracy. They were able to calculate corrections to the energy levels of the hydrogen atom to one part in $10^6$. They were able to explain the 0.1% discrepancy between the experimental value of the electron's magnetic moment and the one predicted by Dirac theory. They predicted that the magnetic moment of the electron, $\mu$, would be equal to $\mu(\text{theo}) = eh/4\pi m_e c$ (1.001159655 ± 3) whereas the measured value is $\mu(\text{exp}) = eh/4\pi m_e c$ (1.001159657 ± 4). The two numbers agree to 1 part in $10^9$. The calculation of the electron's magnetic moment by quantum electrodynamics represents the most accurate determination of a number in atomic physics.

We could never come to understand the intricacy of quantum electrodynamics without delving into the mathematics of the theory. We can, however, gain an understanding of the physics behind the theory and at the same time gain a deeper understanding of the nature of the static electric force. Perhaps the most mysterious aspect of the electric force is the concept of action at a distance. It is very hard to conceive how two charged particles separated by a finite distance can interact with each other with no medium between them. A hint of how this can take place, however, is provided by the production and detection of light or photons. Let us consider the process whereby we see the light produced by a lamp. Light is produced within the filament of the light bulb by exciting the atoms of the filament. The atoms are excited by collisions with the electrons passing through the filament as an electric current. The photons are actually emitted by electrons jumping from one orbit to another within their respective atoms. The photons or light travels through empty space as small packets or bundles of energy. The stream of photons enters our eyes and is absorbed by the electrons in our retina. As the electrons absorb photons, they also absorb energy, which causes them to move about. The motion of these electrons is converted by the optical nerve into an electrical impulse, which is transmitted to our brain where we become aware of the light.

What can we learn from this description of the production and the detection of light? What I find most interesting is that the moving electrons in the light filament caused electrons in the retina to move also through the exchange of light or photons between them. The photons were the medium that allowed the electrons in the filament to move the electrons in the retina. In a sense, the photons transmitted a force from the filament electrons to the electrons of the retina.

Perhaps we can apply this lesson to help explain how the electric force between static charges is transmitted through the exchange of virtual photons. At first this seems to violate the principle of the conservation of energy. If one simply claims that the electric force arises from the exchange of photons then one must account for the energy necessary to create the exchanged photon.

In the case of the light produced by the moving electrons in the light fixture the energy necessary to create the photons came from the de-excitation of the atoms of the filament of the light fixture. Those photons that were transmitted to the electrons in the retina transferred their energy to those retinal electrons providing them with the kinetic energy of their motion.

In the case of static charges, however, the particles remain at rest before and after the transmission of the force. The electrostatic force, however, is independent of the motion of the charges. If the charges remain static throughout the electrostatic interaction there is no way to provide the energy for any photon that may be transmitted between the two charges. This is best seen by considering the time just prior to the hypothetical exchange of a photon, the time during the exchange and the time just after the exchange. Just before the exchange and just after the exchange the total energy of the system is just equal to the energy of the two static electrons. But during the time of the transmission of the photon there is, in addition to the energy of the two static electrons, the energy of the exchanged photon, and hence, during this time there is a violation of energy conservation.

This would seem to kill our hypothesis. However, if we invoke the uncertainty principle it is possible to resurrect our scheme. A violation of the conservation of energy is meaningless if it cannot be measured. If there is no way of detecting the photon during the time of its exchange then we do not have to worry about the conservation of energy since the total energy before and after the exchange remains the same. We, therefore, shall assume that the static electric force arises from the exchange of a "virtual photon" between two charged particles. We call the photon a "virtual photon" because it cannot be detected directly.

Let us investigate whether the exchange of a virtual photon is inconsistent with conservation of energy and the uncertainty principle. Unfortunately, this investigation will require some rudimentary mathematics, which should not disturb most readers. For those who wish to skip the mathematics, what we shall essentially show is that the uncertainty principle does not permit an observation of the exchanged photon and hence, the violation of energy conservation. We shall also show that the only form of the force consistent with the hypothesis of virtual photon exchange, the uncertainty principle and conservation of energy is the inverse square law.

The uncertainty principle has two forms; one of which states that the product of the uncertainty in measuring the position and the momentum must be greater than h, Planck's constant or $\Delta p \, \Delta x \geq h$. The other form of the uncertainty principle states that the product of the uncertainties in measuring the lifetime and energy of a system is greater than h or $\Delta E \, \Delta t \geq h$. It is the latter form that we shall exploit now. The system of the virtual photon will have a lifetime, $t = R/c$ since this is the time required for the photon to propagate the distance, R, between the two charges. We assume the energy of the virtual photon is related to the potential energy of the two charged particles, which in turn depends on the force. The force, we learned, is inversely proportional to the square of the distance between the two charges.

If the two particles are electrons then the force is given by, $F = ke^2/R^2$ where k is a constant that we can take equal to one. The energy of the virtual photon is its potential energy, $E_p$, that is related to the force, F, by the expression, $E_p = FR$ and hence $E_p = e^2/R$. The lifetime of the virtual photon, t, is the time it takes the virtual photon to travel the distance R between the two charges and equals R/c. The product of the energy of the photon, $E = E_p = e^2/R$ times its lifetime $t = R/c$ is just given by $Et = e^2/R \times R/c = e^2/c$.

The quantity $e^2/c$ has the same dimensions as h and just happens to be equal to $h/137$. Since the product of the quantities that we wish to measure is 137 times the product of the uncertainty of measuring these quantities it is quite evident that it will be impossible to detect the virtual photon, and hence, our hypothesis of virtual photon exchange is not inconsistent with the law of conservation of energy.

It is interesting to note that the distance R between the two charges dropped out of the expression for Et. This result only occurs for the inverse square law. If the force had any other form the expression for E × t would have depended on R. It would then be possible to find a value of R for which Et became greater than h and hence observable in principle. If this were to occur then we would have to give up the notion of virtual photon exchange and look for another explanation of action at a distance. For example, if the force were $e^2/R^3$ then the potential energy would be $e^2/2R^2$ and the product of energy times lifetime would be Et = $e^2/2R^2$ × R/c = $e^2/2Rc$ = h/274R. By considering R small enough it would be easy to make Et greater than h and hence, the photon detectable.

The hypothesis of virtual photon exchange gives us some understanding of why the electric force goes like the inverse square of the distance. It is the only dependence on the distance R that is consistent with the uncertainty principle and the conservation of energy. The uncertainty principle plays an extremely interesting role in the virtual photon hypothesis. Rather than limiting our knowledge of the universe the uncertainty principle enables us to understand certain aspects of basic physics more deeply, namely the uniqueness of the inverse square law of the electric force.

The hypothesis of virtual photon exchanges explains how action at a distance occurs. Electrons interact through the exchange of virtual photons. Individual electrons are continuously emitting and reabsorbing virtual photons. They are surrounded by a cloud of virtual photons. Whenever another charged particle enters that cloud it will interact electrically (i.e. experience a force) by absorbing one of the virtual photons. The electron whose virtual photon is absorbed will also experience a force because the absorbed virtual photon will never return to balance the momentum lost by its original emission.

Quantum electrodynamics is based on the hypothesis of virtual photon exchange. Dirac's theory is based essentially on the exchange of a single photon between charged particles. Quantum electrodynamics provides a more detailed description of charged particles by taking into account the exchange of more than one photon. Multiple photon exchange can be represented pictorially with diagrams first devised by Richard Feynman. Fig. 22.1 shows the diagram for single photon exchange.

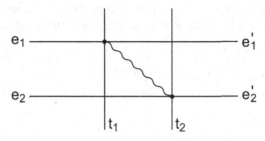

Fig. 22.1 Photon exchange by electrons.

Time is increasing in this diagram as we move from left to right. The lines labeled $e_1$, $e_2$, $e'_1$ and $e'_2$ represent electron 1 and 2 before and after their mutual interaction. At time $t_1$, $e_1$ emits a virtual photon represented by the wiggly line, which is absorbed by $e_2$ at time $t_2$. Note that as a result of the emission at $t_1$, $e_1$ changes to $e'_1$ and as a result of the absorption at $t_2$, $e_2$ changes to $e'_2$. Corrections to this diagram are shown below in which two photons are exchanged. There are three possible diagrams of this type shown below in Fig. 22.2.

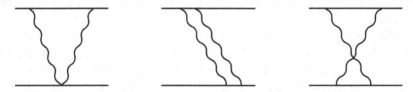

Fig. 22.2 Two photon exchange diagrams.

The first two diagrams are equivalent to each other so that there are only two types of diagrams involving the exchange of two photons. The contribution of the two-photon exchange diagrams of Fig. 22.2 is a small correction to the one photon exchange diagram of Fig. 22.1. Each time a photon is exchanged a factor of $e^2/hc$ enters the probability of the exchange taking place. Since $e^2/hc = 1/137$ the relative probability of two photon exchanges to one photon exchange is $1/137$. The relative probability of n photon exchange to one photon exchange is $(1/137)^{n-1}$. Each successive order of photon exchange becomes less and less probable. Nevertheless, these small corrections have been calculated and have been shown to agree with experiment as pointed out above. In addition to the diagrams in which two photons are exchanged between the charged particles, there are other two-photon diagrams in which only one photon is exchanged. We give three examples of these types of diagrams, Fig. 22.3:

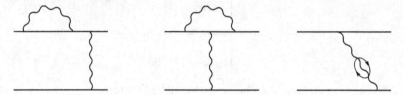

Fig. 22.3 Two photon diagrams where only a single photon is exchanged.

In the first diagram (Fig. 22.3) the electron emits and absorbs a photon before exchanging the photon with the other charged particle. This type of diagram does change the nature of the electric interaction but it does not change the mass of the charged particle. The second diagram also does not change the nature of the electric force but it does affect the strength of the particle's charge. The third diagram involves the creation of a virtual electron-antielectron pair from the virtual photon. This diagram gives rise to a correction of the electron's magnetic moment.

We shall conclude our discussion of quantum electrodynamics by presenting the Feynman diagrams representing pair creation and pair annihilation. For pair creation, the presence of a nucleus is necessary in order to conserve energy and momentum. We shall represent the nucleus by the symbol Z in our diagram, Fig. 22.4, and the scattered nucleus by Z'. In the process of creation we see a real photon and a nucleus entering.

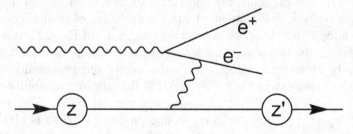

Fig. 22.4 Electron-positron pair creation reaction.

The real photon turns into an electron positron pair. The electron then exchanges a virtual photon with the nucleus.

Pair annihilation occurs in the following diagram, Fig. 22.5 through the spontaneous emission of a photon by the electron followed by the annihilation of the electron and positron into the second photon:

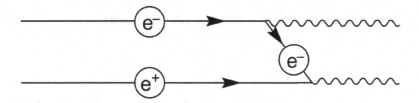

Fig. 22.5 Electron-positron pair annihilation.

One can look on this diagram as the e⁻ exchanging a virtual e⁻ and turning into a photon. The positron then absorbs the virtual e⁻ and also turns into a photon. The last diagram has an interesting interpretation if one assumes that a positron is an electron moving backwards in time. Then the pair annihilation has the form where it would appear the electron enters the diagram, Fig. 22.6, at the top and changes its direction in time and exits the diagram at the bottom.

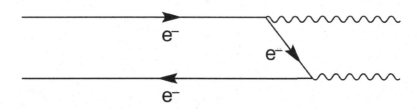

Fig. 22.6 Electron-positron pair annihilation in which the positron is represented as an electron moving backwards in time.

# Chapter 23

# The Nucleus and the Strong Interaction

So far in our study of the atom we have basically examined the nature of the electromagnetic force and the interactions of electrons and antielectrons with photons. The electromagnetic force, however, cannot explain how the protons and neutrons that make up the nucleus of the atom can be confined to a space of the order of only $10^{-13}$ cm. Because the electric force between the protons is repulsive there must be another force that binds the nucleons, i.e. the protons and neutrons, together in the nucleus of atoms. This force is the nuclear force or the strong interaction, which is approximately 100 times stronger than the electric force.

The atomic nucleus was first discovered in 1912 by Rutherford and his co-workers in the experiment described in Chapter 19 in which atoms were bombarded by $\alpha$ particles. These experiments revealed that the nucleus is less than $10^{-12}$ cm in radius. Later experiments show that this number was closer to 2 or 3 × $10^{-13}$ cm (or 2 or 3 fermi). The fermi (fm) is a unit of length equal to $10^{-13}$ cm used frequently in nuclear physics because just about all distances in this field range from 1 to 10 fermis. Because the electron is such a light particle, the nucleus carries almost all the weight of the atom. The volume the nucleus occupies, on the other hand, is only $10^{-15}$ the total volume of the atom. The density of matter in the nucleus is therefore quite considerable. It is in fact approximately $10^{14}$ times the density of water.

The charges and masses of individual nuclei were discovered basically by the chemists. From this information it became immediately evident that the proton was not the sole constituent of the nucleus. The helium nucleus for example has a mass approximately four times the hydrogen nucleus but a charge of only +2e. Before the discovery of the neutron it was believed that the helium nucleus consisted of 4 protons and 2 electrons. It was only after the discovery of the neutron in 1930 that it was realized that the helium nucleus is composed of two protons and two neutrons. The proton and neutron are referred to generically as nucleons. This term is used to describe the two particles because their interactions within the nucleus are more or less the same with only minor corrections because of the difference of their charges. The mass number of a nucleus, A, is equal to the total number of nucleons. The atomic number, Z, is equal to the total number of protons and hence the total number of electrons. The total number of neutrons in the nucleus, N, equals A-Z. The mass of the nucleus is approximately equal to MA where M is the mass of a nucleon (the mass of the proton and a neutron are approximately equal). The charge of the nucleus is exactly equal to +Ze.

Not all nuclei are stable. Some are radioactive which means that they spontaneously change into another nucleus through the emission of an $\alpha$ particle, an electron, a positron or a neutron. It was through the study of the chemistry of radioactive nuclei that it was discovered that there are a number of nuclei with the same value of Z but a different value of A. Two nuclei with the same number of protons but a different number of neutrons are called isotopes. Most elements consist of a number of isotopes as any inspection of the periodic table reveals. The existence of isotopes explains why the mass of all elements is not an even integer times the mass of hydrogen. The element chlorine has the atomic weight of 35.5 because 80% of the element is $Cl_{35}$ and 20% $Cl_{37}$. Both $Cl_{35}$ and $Cl_{37}$ consist of 17 protons but $Cl_{37}$ has 20 neutrons while $Cl_{35}$ has only 18. An atom of $Cl_{35}$ and an atom of $Cl_{37}$ behave identically from a chemical point of view despite the fact that one is heavier than the other. The same is true of uranium-235 and uranium-238; they are identical chemically. Their nuclei behave quite differently, however. The lighter isotope is much more radioactive than the heavier one. $U_{235}$ participates much more readily in the process of nuclear fission than does $U_{238}$.

## The Nuclear Force

The very first question that the existence of nuclei raises is the nature of their stability. The protons of the nucleus all have the same charge and hence electrically repel each other. The strength of this repulsion is quite strong since the protons are so close to each other. The electric force between two protons at a distance of one fermi from each other equals 250 Newtons. This force acting on a free proton would produce an initial acceleration of over $10^{29}$ m/sec$^2$ or $10^{28}$ times the acceleration of a body falling at the surface of the earth. How is the stability of the nucleus to be explained in terms of this enormously strong repulsive force? The neutron carries no charge and therefore cannot contribute any attractive force. The gravitational force is attractive but much, much weaker than the electric force by a factor of $10^{40}$ and hence can be completely ignored. The magnetic forces, which arise from the magnetic moments of the proton and the neutron are also much weaker than the electric force and are not necessarily always attractive. The only way we can account for the stability of nuclei, which, implies a net attractive force between all the nucleons is to assume the existence of another force much stronger than the electric force, which we shall appropriately call the nuclear force.

The nuclear force is also referred to as the strong interaction because of its great strength. The nucleus force is the same for neutrons as it is for protons. The force is a short-range force, which acts only when the separation of the nucleons is less than 1 or 2 fermis. The force in this range is extremely strong about 100 times the electric force and basically attractive. If the separation of the nucleons becomes really small, however, less than ½ a fermi, then the force becomes repulsive. The size of the nucleus is determined by the nature of the nuclear force. The radius of nucleus can never be more than a few fermis otherwise there would be no attractive force to hold it together. It also cannot be much less than a fermi, on the other hand, because the nuclear force becomes repulsive at really small distances preventing a general collapse of the nucleus. Much has been learned about the nuclear force by studying the scattering of protons by protons. We shall discuss this interaction in greater detail in the next chapter on elementary particles. For the purpose of discussing nuclear physics the above description of the nuclear force will suffice.

## Nuclear Binding Energy

One of the earliest indications of the strength of the nuclear force was the size of the nuclear binding energy compared with atomic binding energy. If two particles form a bound state due to an attractive force, then the total mass of the bound state is less than the sum of the masses of the individual particles composing the system. The reason for this is that the potential energy of the bound particles is negative. If one wished to separate the two particles, one would have to supply energy to do work against the attractive force, hence the negative potential energy. The rest mass energy of the bound state is less than that of the sum of the rest mass energies of the individual components making up the bound state because of the negative potential energy and because as Einstein discovered mass and energy are equivalent. In fact the attenuation of nuclear mass due to binding energy formed one of the basic tests of Einstein's hypothesis of the equivalence of mass and energy.

The binding energy of the electron in the hydrogen atom is only 13.6 electron volts (eV). An eV is a unit of energy equal to the energy an electron gains when accelerated through a one volt potential. The eV is the commonest unit of energy in nuclear and elementary particle physics. Rest mass energies are also measured in electron volts or eVs. Since mass and energy are equivalent, one can specify the mass of a particle by specifying the energy of the particle's rest mass, which is equal to $mc^2$ and is measured in eVs or MeVs (million electron volts). The rest mass energy of the electron is approximately 0.5 MeV and the rest mass energy of the proton approximately 940 MeV. The rest mass energy of the hydrogen atom is 13.6 eV less than the sum of the rest mass energies of the proton and the electron. The binding energy hardly affects the mass of the hydrogen atom. Nuclear binding energies are much greater than atomic binding energies and involve rest mass energies in the MeV range.

The reason the nuclear binding energies are greater is because the nuclear force is stronger than the electric force and also the separation of the particles involved is much smaller. Let us consider the binding energy of one of the simplest nuclei known, the deuteron. The deuteron, an isotope of hydrogen, is a bound state of a proton and a neutron. The difference in mass between the deuteron and the sum of the masses of the proton and the neutron is just equal to the nuclear binding energy of the deuteron. The rest mass energy of the proton is 938.27 MeV, of the neutron 939.56 MeV and of the deuteron 1877.7 MeV. The binding energy of the deuteron in MeV is therefore 938.27 + 939.56 − 1875.61 = 2.22 MeV.

If we would want to separate the neutron and the proton in the deuteron we would be obliged to supply an energy of at least 2.22 MeV.

This can be tested by studying the interaction of photons with deuterons. If the energy of the photon is not very great, the deuteron will scatter the photon. If the energy of the photon becomes large enough, however, the photon is capable of destroying the nuclear bond and disintegrating the deuteron. The process of the photo-disintegration of the deuteron may be represented as follows: $\gamma + D \rightarrow p + n$. The threshold energy for which this reaction occurs is exactly 2.22 MeV, the binding energy of the deuteron. Unless the photon provides enough energy to provide the necessary mass, the deuteron will not disintegrate.

The nuclear binding energy for other nuclei is determined in exactly the same way we determined it for the deuteron. Let us consider an arbitrary nucleus of Z proton and N neutrons with a mass M. The binding energy of the nucleus, $E_B$, is obtained by subtracting the mass of the nucleus from the mass of the nucleons and hence $E_B = Zm_pc^2 + Nm_nc^2 - Mc^2$.

The binding energy turns out to be basically proportional to the number of nucleons $A = Z + N$. This indicates that the nucleons inside a nucleus do not interact with all the other nucleons but rather the nucleus force is saturated and each nucleon only interacts with two or three other nucleons in the nucleus. If the nucleons were interacting with all the nucleons in the nucleus, there would be $A(A - 1)/2$ individual nucleon-nucleon interactions among the A nucleons in the nucleus. If the potential energy of each nucleon-nucleon interaction were equal to $E_{NN}$, then one would expect the binding energy $E_B$ to equal $A(A - 1) E_{NN} /2$. One would then find the binding energy proportional to $A^2$ and not A. The saturation of the nuclear force indicated by the behaviour of $E_B$ is another reflection of the finite range of the nuclear force.

The neutron and the proton have the same nuclear force. This fact is reflected in the fact that the number of neutrons and protons in the stable nuclei are about the same. The reason that stable nuclei composed exclusively or principally of either neutrons or protons does not occur is because of the Pauli exclusion principle. The proton and the neutron are both spin ½ particles and hence are affected by the Pauli exclusion principle, which does not allow more than one proton or one neutron to have the same quantum numbers. A proton and a neutron may have the same quantum numbers in the nucleus because they are different particles. The nucleons in a nucleus assume different energy states somewhat like the electrons surrounding the nucleus. If a nucleus were filled exclusively with protons, the higher energy states would become filled quickly. If I can fill the states with both neutrons and protons, I can put a neutron and a proton into each energy state and therefore do not fill the higher energy states as fast. The lower the energy of all the nucleons, the more binding energy there is and hence the more stable the nucleus. This explains the near equality of neutrons and protons in the stable nuclei. The ratio of neutrons to total nucleons increases from 50% for the light nuclei to about 60% for the heaviest nuclei. This increase is due to the repulsive electric force between protons. The reason for the increase of this ratio is that it costs more energy to add a proton to a heavier nucleus than a neutron because of the repulsive electric force between the proton and the rest of the nucleus.

There are both stable and unstable nuclei. A stable nucleus is one which left undisturbed will remain in the same state. A stable nucleus by definition is in its ground state. Like the atom, the nucleus has various excited states. When a nucleus is in one of its excited states, it will emit electromagnetic radiation making transitions to lower levels until it reaches its ground state. This is familiar to us from our study of atomic physics but differs from the atomic case in that the nuclear radiation is of a much higher energy and hence involves $\gamma$-rays rather than x-rays or visible light.

All nuclei in excited states are unstable and hence radioactive. There are nuclei, however, which are radioactive even in their ground state. These nuclei emit other forms of radiation such as alpha particles (helium nuclei), beta rays (electrons), positrons, protons or neutrons. The radioactive nucleus is transformed into another nucleus as a result of this transition. If the nucleus into which the radioactive nucleus is transformed is the ground state of a stable nucleus, then there will be no more emission of radiation. But if the nucleus is transformed into an excited state it will continue to emit radiation until the ground state of a stable nucleus is reached. The lifetime of radioactive nuclei, which is the time from the creation of the nucleus to its spontaneous decay, varies from nucleus to nucleus. Berylium-8 only lives $3 \times 10^{-16}$ seconds on the average whereas the average lifetime of Lead-204 is $1.4 \times 10^{17}$ years, a time greater than the apparent lifetime of the universe. If the lifetime of a nucleus is $\tau$, it does not mean that all the nuclei decay exactly after time $\tau$ has passed. The decay of each individual nucleus is a random process but the average the lifetime it $\tau$.

Alpha radiation or the spontaneous emission of a helium nucleus is the usual mode of radioactivity of the heavier nuclei such as lead, radium, thorium and uranium. Beryllium-8, however, also undergoes alpha decay, decaying into two alphas. When a nucleus of A nucleons and Z protons emits an alpha particle, it becomes a nucleus with $A - 4$ nucleons and $Z - 2$ protons. The mass of the alpha particle and the daughter nucleus is less than the original mass of the parent nucleus. The mass which is lost in this transition is converted into the kinetic energy of the daughter nucleus and the alpha particle according to Einstein formulae, $E = mc^2$. The kinetic energy of the alpha particles is usually about 5 or 6 MeV.

Beta decay or the emission of an electron occurs more frequently among the lighter radioactive nuclei. In this process a neutron in the nucleus changes into a proton so the parent nucleus of Z protons and N neutrons changes into a daughter nucleus of $Z + 1$ protons and $N - 1$ neutrons. This decay involves a basic force of nature, which we have not yet encountered and which is referred to as the weak interaction. The three other basic forces of nature that we have encountered are the gravitational, electromagnetic and nuclear interactions. The weak decay of a nucleus actually involves the emission of a second particle in addition to the electron. This particle, known as the neutrino, is very unusual. It has very little mass and no charge but carries away energy and momentum and in addition carries a half unit of spin. We shall study this process in greater detail in the next chapter, which deals with the interactions of elementary particles. Not only do the neutrons in unstable nuclei experience beta decay, but also the free neutron decays in this manner. The average lifetime of a neutron is 1000 seconds after which it decays spontaneously into a proton, an electron and a neutrino.

$$\text{Neutron} \rightarrow \text{proton} + \text{electron} + \text{neutrino } (n \rightarrow p + e + \nu)$$

We shall leave the discussion of this decay for the next chapter. Closely related to the process of electron emission is positron emission, which also involves the weak interaction. In this process a proton in an unstable nucleus changes into a neutron plus a positron (an antielectron) plus a neutrino. This means that a nucleus with Z protons and N neutrons as a result of positron emission is converted into a nucleus with $Z - 1$ protons and $N + 1$ neutrons. A free proton does not undergo positron emission because the mass of the neutron, the positron and the neutrino are greater than that of the proton and hence the proton does not have the energy to make this transition. For all the radioactive nuclei, which emit electrons or positrons, the mass of the parent nucleus is always greater than the mass of the daughter nucleus plus the mass of the electron (or positron) and the neutrino. The mass which is lost is converted into the kinetic energy of the daughter nucleus, the electron (or positron) and the neutrino according to Einstein's formula $E = mc^2$.

A great deal has been learned about nuclei by studying the products of their radioactive decays. Much more has been learned, however, by studying the effects of bombarding the nucleus with various projectiles such protons, neutrons, electrons, photons, alpha particles and other nuclei. There are two possible outcomes of subjecting the nucleus to collisions with other particles. Either the particles will scatter off the nucleus without changing their character or else they will initiate nuclear reactions in which the incoming projectile and the nucleus undergo change.

The very first collision experiments were performed using radioactive materials as a source of alpha particles, which bombarded the nuclei of the target, usually a thin-foil of some metal such as gold. Rutherford's discovery of the nucleus described earlier was the first such experiment. As experimental techniques became more sophisticated, nuclear physicists designed machines that accelerated charge particles through electric potentials before directing them at a target. The first machine of this type was the Van De Graff electrostatic generator. The large electrostatic potential through which the protons were accelerated was generated using friction. The energy of the particle that one obtained in this manner was limited by the amount of potential difference the machine could be made to hold. The next innovation in accelerators avoided the problem of electric breakdown by recycling the charged particle through the same potential difference many times. This was achieved by forcing it into a circular orbit through the use of magnets. The more sophisticated accelerators developed since the first cyclotron are still based on this principle.

The detection of the products of a scattering experiment posed another challenge to the ingenuity of the nuclear physicist. Rutherford detected the alpha particles scattered from the gold foil using a material, which scintillates each time it is struck by a charged particle. This simple technique still forms the basis for many of today's particle detectors. The signals from the scintillating material are now automatically read by a photoelectric cell whose signals are fed directly into a computer, which analyzes the data, almost the instant it is gathered. Another device for observing nuclear reactions is the cloud chamber. In recent times the cloud chamber has been replaced by the bubble chamber, which operates on a similar principle. In the cloud chamber the charged particle leaves tracks by condensing droplets of water in the vapor, whereas in the bubble chamber tracks are left because the charged particles vaporize gas bubbles in the superheated liquid of the bubble chamber. The development of accelerators and particle detectors has allowed the nuclear physicist to peer into the nucleus to discover its secrets.

When I was a graduate student one of my professors likened the study of the nucleus through scattering experiments to the situation of a man trying to discover the nature of an avocado in the following bizarre fashion. Somebody places an avocado in a pitch-black room. The physicist enters the room blindfolded with a machine gun and fires a hail of bullets in the direction of the avocado. The physicist leaves the room and the avocado is removed. The physicist is allowed to reenter the room and turn on the lights. By studying the pieces of avocado on the wall he or she is then expected to determine the nature of the fruit they fired at, the size of the fruit, the size of the pit, the thickness of the skin, etc. This is the task, which faces nuclear physicists. They fire into the dark, never seeing what they are aiming at. Hopefully, by detecting bits of matter that emerge from the interaction of the projectile and the nucleus, they will discover the nature of the nucleus and the nuclear force.

The simplest bombardment process of the nucleus is scattering in which the incoming particle is deflected by the nuclear or electric force of the nucleus. If the incoming particle is an electron, it will only be deflected by the electric force since the electron has an electric charge but no nuclear charge. The electron is unaffected by the nuclear force. Electron scattering experiments have revealed the charge distribution of the nucleus. Scattering experiments in which the proton and the neutron are the incoming projectiles have yielded information on the nature of the nuclear force such as its dependence on the position of the particles and the alignment of their spins. These scattering experiments have also helped to determine the size and shape of individual nuclei.

In most scattering experiments the energy of the projectile before and after the collision is the same as is the energy of the nucleus. In some cases the nucleus is left in an excited state as a result of the collision. The incoming particle suffers an energy loss just equal to the energy gained by the nucleus. The nucleus does not remain in its excited state for very long but quickly decays into its ground state by emitting gamma rays. From the study of the inelastic scattering just described one is able to discover the energy levels of the nucleus just as one studies the energy levels of the atom through atomic spectroscopy.

Nuclear reactions in which the incoming projectile and target nucleus completely change their character also provide an opportunity to study the energy levels of the nucleus among other things. The first nuclear reaction was observed in 1919 by Rutherford when he bombarded nitrogen nuclei with alpha particles obtained from a radioactive source. He found that the helium and nitrogen nuclei were transformed into a proton and the oxygen-17 nucleus: $He^4 + N^{14} \rightarrow O^{17} + p$.

By initiating this reaction Rutherford had realized the alchemist's dream of artificially transmuting a basic element. He had transformed nitrogen into oxygen using a radioactive source. In 1932 Cockcroft and Walton were the first to induce the transmutation of an element by bombarding a nucleus with an artificially accelerated particle. They directed a beam of protons at Lithium-7 nuclei. They found that the Li-7 absorbed the proton to form Beryllium-8 and quickly broke up into two alpha particles: $p + Li^7 \rightarrow He^4 + He^4$.

In 1936 Niels Bohr explained the mechanism of low energy nuclear reactions as a two-step process. The first step involved the formation of a compound nucleus composed of all the nucleons in the reaction. The compound nucleus lives a very short time, approximately $10^{-18}$ sec, and then decays into the final state. The relative probability of the states into which the compound nucleus decays is independent of how it is formed.

Let us consider, for example, the production of the excited state of nitrogen-14 ($N^{14*}$). This state can be formed by bombarding carbon-13 nuclei with protons. The compound nucleus $N_{14}^*$ can decay into a number of final states:

$$p + C^{13} \rightarrow N^{14*} \rightarrow B^{10} + He^4 \text{ or } C^{12} + H^2 \text{ or } N^{13} + n \text{ or } N^{14} + \gamma.$$

The relative probabilities of these final states would be the same if we had created $N^{14*}$ by bombarding boron-10 nuclei with alpha particles or nitrogen-13 nuclei with neutrons.

The energy levels of a nucleus are determined by measuring the cross section for a nuclear reaction as a function of the energy. The cross section for a reaction is the probability that the reaction will take place. If the energy of the incident particle plus the rest mass energy of the nucleus correspond to the energy of an excited state of the compound nucleus, the probability of the reaction taking place will be large, and hence there will be a peak in the cross section. The peaks of the cross section correspond to the energy levels of the excited compound nucleus. The widths of the peak represent the uncertainty in measuring the energy of the excited state of the nucleus.

According to the uncertainty principle, the uncertainty in the energy times the uncertainty in the time of a system is equal to h. The uncertainty in time of the compound nucleus just corresponds to its lifetime. We therefore expect on the basis of the uncertainty principle that the lifetime of the compound nucleus is approximately h/ΔE. The widths of the experimentally measured peaks of the cross section, therefore, determine the lifetimes of the compound states. Knowing the energy and lifetime of the excited states of the nucleus helps physicists determine the structure of the nucleus.

One of the factors determining the final states of a nuclear reaction is the conservation of energy. There are two types of nuclear reactions, depending on whether the masses of the final state nuclei are heavier or lighter than the initial state nuclei. When the final state is heavier, the reaction takes place by converting a certain amount of the initial kinetic energy of the incident particle into rest mass energy. An example is the reaction $p(1.0078) + C^{13}(13.0034) \rightarrow H^2(2.0141) + C^{12}(12.00)$, where the numbers in parentheses are the mass of the nuclei in atomic mass units (amu). The total mass of the initial nuclei is 14.0112 whereas the final state masses sum to 14.014 and hence are 0.0029 amu heavier than the initial state masses. This reaction has a threshold energy because unless the proton has a certain amount of kinetic energy, which can be converted into rest mass energy, the reaction will not proceed.

In the other type of reaction the final masses are lighter than the initial masses and hence rest mass energy is converted into kinetic energy. These reactions have no threshold since there is enough energy for the reaction to proceed in the rest masses of the initial state. There are two types of reactions which release nuclear energy. The first type is the fission reaction in which a heavy nucleus divides into two or more parts as a result of absorbing a neutron. The second type is the fusion reaction in which two light nuclei combine together to form a heavier nucleus. Whether fission or fusion is possible depends on the binding energy per nucleon, which increases with A until a maximum is reached at A = 60. If two nuclei in the region A < 60 combine to form a third nucleus, the total binding energy will increase and hence the heavier nucleus will have less mass than the two nuclei which combined to form it and fusion will take with the release of energy as occurs when 4 protons combine to form a helium nucleus.

After the maximum at $A = 60$, the binding energy per nucleus decreases. If a nucleus with $A > 60$ were to split into two medium sized nuclei, then the total binding energy would increase and the total mass would decrease. The destruction of mass would then result in the release of energy as occurs with nuclear fission

The first fission reaction was discovered when uranium was bombarded by neutrons. It was discovered that the uranium nucleus divided into barium and krypton plus three additional neutrons. The release of the additional neutrons allows for the possibility of a chain reaction since the extra neutrons can trigger the fission of other uranium nuclei which in turn release extra neutrons, etc. the fission reaction involves the release of 200 MeV per nucleus and requires only $10^{-14}$ sec. Because the chain reaction spreads so quickly, a great deal of energy is released in a very short time. This is the principle behind the release of energy in the atomic bomb. It is possible to slow down the rate at which new fission reactions are triggered by not concentrating the uranium in one lump and by placing material, which absorbs neutrons between the various concentrations of uranium. By controlling the amount of neutron insulating material, the rate of fission is controlled allowing the amount of atomic energy released at a given time to be controlled, which is how a nuclear power plant works. The term atomic energy is really a misnomer since the energy released in fission is nuclear energy and not atomic energy. Atomic energy is the energy released in chemical reactions.

Fusion reactions always involve the release of energy, however a certain amount of kinetic energy is needed at the outset in order for the reactions to proceed. A neutron can penetrate a heavy nucleus at any energy because it is neutral and does not have to overcome a repulsive electric force. If two nuclei try to combine together, they are pushed apart by their respective positive charges. There is, therefore, a threshold energy for fusion reactions below which the nuclei will not combine. Once the threshold is exceeded, however, nuclear energy is readily released. The fusion of hydrogen into helium is responsible for the release of energy in the hydrogen bomb. In order to achieve the threshold energy necessary to make this reaction proceed, an atomic bomb is exploded in the heart of the hydrogen bomb. The high temperature generated by this explosion then triggers the fusion reaction.

The fusion reaction is also responsible for the release of energy in the stars and the Sun. The threshold energy for fusion in a star is easily achieved because of the high temperatures of the star. A star before it is ignited by fusion achieves the threshold energy for fusion through gravitational collapse. Once enough interstellar material collects (basically hydrogen gas), a gravitational collapse occurs in which extremely high temperatures are reached. Eventually the kinetic energy of the nuclei becomes so great that the hydrogen fuses into helium. The process is actually quite complicated, proceeding through several steps: First, two hydrogen nuclei fuse to form a deuteron $H^1 + H^1 \rightarrow H^2 + e^+ + \gamma$. Then another proton combines with the deuteron to form helium-3, $H^1 + H^2 \rightarrow He^3 + \gamma$. Finally, two $He^3$ nuclei combine to form $He^4$ and two protons, $H_e^3 + H_e^3 \rightarrow H_e^4 + 2 H^1$.

# Chapter 24

# Elementary Particles, Quarks and Quantum Chromodynamics

In this chapter we will examine the nature of elementary particles. But before doing that we need to note that the definition of what is an elementary particle has changed in recent years. In the days that chemists and physicists discovered that the basic units of chemistry and the structure of gases, liquids and solids were atoms and molecules it was the atoms that were considered to be the most fundamental particles of the universe and that molecules were composites of atoms. The term atom is derived from the ancient Greek term ἄτομος (átomos) meaning not divisible. As it turned out the atom was divisible and was shown to be composed of protons, neutrons and electrons, which were thought to be the smallest possible particles and were dubbed the elementary particles. The neutrino, which first appeared in the beta decay of neutrons, was added to this category.

As we will soon see the number of elementary particles that were discovered exploded and that many of these particles including the proton, the neutron and many other particles including mesons and baryons that interact through the nuclear force and are known as hadrons were shown to be composites of still more elementary particles called quarks and antiquarks. The electron and neutrino and other members of their class known as leptons are not made of quarks and they retain their designation as elementary particles. In addition to the quarks and leptons there are a small number of bosons that give rise to the fundamental forces of electromagnetism, the nuclear force or strong interaction, and the weak interaction. These bosons include the photon, which gives rise to electromagnetism, the gluon, which gives rise to the strong interaction and the W and Z bosons, which give rise to the weak interaction. Strictly speaking the class of elementary particles consists of the quarks, the leptons and the bosons. The hadrons like the proton (p) and the neutron (n) are composites of quarks and strictly speaking are no longer elementary but they are the subject of elementary particle physics and are regarded by some as elementary particles.

Elementary particle physics comprises the study of leptons and hadrons composed of quarks and their strong, electromagnetic and weak interactions mediated by the bosons. The only other force in nature is gravity, which is studied in the domain of general relativity and cosmology. Attempts to integrate gravity with the three basic forces of the strong, electromagnetic and weal interactions have met with little or no success.

The electric and magnetic forces, as well as the interactions of light, were shown by the work of Maxwell and Einstein to be the result of a single electromagnetic interaction. They arise as a result of the exchange of virtual photons. The nuclear force due the exchange of mesons has a range of $10^{-13}$ cm, is the same for neutrons and protons and is approximately 100 times the strength of the electric force.

## The Nuclear Force or the Strong Interaction

The first attempts to explain the nuclear force took their inspiration from the fact that the electromagnetic interaction between charged particles arises from the exchange of a virtual photon. Because the photon is massless, the range of the electromagnetic force is infinite since it is always possible to choose a photon with an energy small enough such that the product of its energy with time for its exchange is less than Planck's constant, h, and hence unobservable. With a massive particle, however, the minimum energy of the exchanged particle is at least equal to its rest mass energy, $mc^2$. The minimum time for transit of the exchanged particle is t = R/c where R is the distance between the interacting particles and c is the maximum allowed velocity for the exchanged particle. If the exchanged particle is to remain unobservable within the context of the uncertainty principle, then the product of its energy, $mc^2$, with its transit time, R/c must be less than or equal to h, or $mc^2 \times R/c \leq h$ so that the range of the force is approximately R = h/mc.

The range of the interaction is inversely proportional to the rest mass of the exchanged particle. We, therefore, see immediately why the range of the electromagnetic force is infinite since the rest mass of the exchanged particle, the photon, is zero. The first to attempt an explanation of the nuclear force due to the exchange of virtual particles was Heisenberg in 1934, who thought that a neutron and a proton could have a strong interaction by the exchange of an electron and a neutrino. But the range $R = h/m_e c$ of this interaction would be $4 \times 10^{-11}$ cm, which is about 400 times the actual range of the nuclear force. This hypothesis had to be abandoned.

A year later, in 1935, the Japanese physicist H. Yukawa proposed a theory for the strong interaction incorporating Heisenberg's idea of exchange. He postulated that nucleons interact by the virtual exchange of a new particle which he called a mesotron and which later was shortened simply to meson. Yukawa assumed that this particle was strongly coupled to nucleons and that it had a rest mass somewhere between 100 and 200 MeV. The choice of the meson's mass was made in order to explain the range of the nuclear force. The name meson comes from the Greek word for middle and refers to the fact that the meson has a mass between that of the electron and the proton.

The particle was assumed to have an electrically neutral state as well as two charged states with charge ±e. This enables one to explain both the direct force and the exchange force. The neutron and proton exert a direct force on each other by the exchange of a neutral meson and an exchange force by the exchange of a charged meson.

Within a year of Yukawa's predictions, Anderson, the discoverer of the positron, observed a new particle among the cosmic ray tracks he was studying. The new particle had a mass of 105 MeV, which was right within the range predicted by Yukawa. This discovery generated a great deal of excitement since it seemed to confirm Yukawa's hypothesis. Then in 1947 the bubble burst when Conversi, Pancini, and Piccioni through their cosmic ray work discovered that this new particle had a very weak coupling to the proton and neutron and hence could not be the intermediate meson predicted by Yukawa. This new particle, which was called a muon and is denoted by $\mu$, is interesting in itself. It behaves to all extents and purposes exactly like an electron except that it has a mass 210 times as large as the electron's mass. It has the same charge and spin as the electron. It also has an antiparticle, the positive muon. The muon is unstable and decays into the electron an antielectron neutrino and a mu neutrino via the weak interaction $\mu^- \rightarrow e^- + \nu_\mu + \bar{\nu}_e$.

The muon together with the electron, the neutrino and their antiparticles for the class of particles called leptons after the Greek word for light. There are two types of neutrinos associated with the electron, the $v_e$ and the $v_\mu$. We shall return to the muon and the other leptons when we will discuss the weak interaction in greater detail.

The disappointment over the lack of interaction of the muon with nucleons was very keen. However, in 1947 another study of cosmic rays by Powell, Lattes, and Occhialini revealed the existence of a second intermediate mass particle, the pion or $\pi$ meson. This new particle had all the features required by Yukawa's theory. It interacts strongly with nucleons, it has three charge states, $\pi^+$, $\pi^0$ and $\pi^-$ and finally its mass of 140 MeV can explain the $10^{-13}$ cm range of the nuclear force.

Our image of the neutron and the proton changes as a result of Yukawa's model of the strong interaction. The proton and neutron can no longer be considered stable particles. They are continuously in flux, changing their state every $10^{-23}$ seconds, perpetually surrounded by a seething, surging cloud of mesons. The neutron and the proton are also continually exchanging roles, each becoming the other by exchanging the appropriately charged $\pi$ meson. For the proton the following transformations take place, $p \rightarrow n + \pi^+ \rightarrow p \rightarrow p + \pi^0 \rightarrow p$. It represents the proton transforming itself into a neutron and a positively charged pion and then back into a proton and then into a proton and a $\pi^0$ and then back again to a proton. These transformations may also be represented by Feynman diagrams like the ones introduced to designate the scattering of charged particles through virtual photon exchange. This transformation would appear in the language of Feynman diagrams as follows:

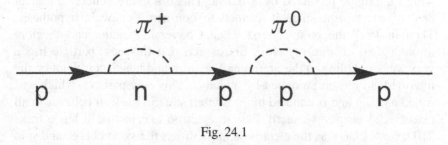

Fig. 24.1

The neutron undergoes a similar transformation turning into a proton and a negatively charged pion and then back into a neutron and then into a neutron and a neutral pion and finally back into a neutron, i.e.,

$$n \rightarrow p + \pi^- \rightarrow n \rightarrow n + \pi^0 \rightarrow n \, .$$

The proton and neutron are continually emitting and reabsorbing mesons. The mesons can never wander more than $10^{-13}$ cm from the proton or else they will no longer be masked by the uncertainty principle and a violation of energy conservation would be detected. If another nucleon wanders into the meson cloud, a meson exchange will take place and the two nucleons will exert a nuclear force upon each other. This is illustrated in Feynman language with the following diagram where nucleon $N_1$ emits a pion at time $t_1$ which is absorbed by nucleon $N_2$ at time $t_2$:

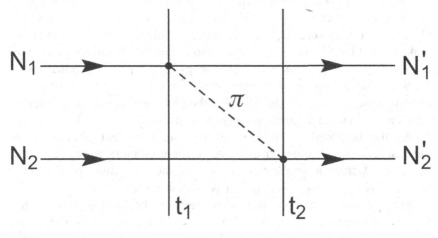

Fig. 24.2

The transformations of the proton can be more complicated since the protons can emit more than one meson at a time in which case a transformation like the following is possible: $p \rightarrow p + \pi^0 + \pi^- + \pi^+ \rightarrow p \rightarrow n + \pi^+ + \pi^0 \rightarrow p$. These kinds of transformations can lead to scattering between nucleons where multiple mesons are being exchanged. The proton and the neutron behave in a completely symmetric way as far as the nuclear force and their interaction with mesons are concerned.

The virtual mesons hovering about the nucleon are unobservable because of the requirements of energy conservation. The proton or neutron seems to borrow energy to produce the meson and must repay that debt before anyone can detect the deficit by reabsorbing the meson or exchanging it with another nucleon, which is within a convenient distance. If the proton is provided with a source of energy, however, there is no reason why the meson would have to be returned. On the basis of Yukawa's theory, therefore, one would expect to produce mesons by providing a nucleon with enough energy to provide for the rest mass energy of the meson. One mechanism for providing this energy is to knock one nucleon against another hard enough so that the kinetic energy of the incoming nucleon can be transformed into the rest mass of the $\pi$ meson, allowing a virtual meson to escape. Yukawa's theory, therefore, predicted that once protons could be accelerated to kinetic energies of 140 MeV or more that the following reactions would be observed: $p + p \rightarrow p + p + \pi^{o}$ or $p + p \rightarrow p + n + \pi^{+}$ or $p + n \rightarrow p + p + \pi$.

Once accelerators were build that could achieve these energies, the above reactions were indeed observed lending greater credence to Yukawa's theory. Mesons were also released from nucleons by bombarding the nucleons with high-energy photons so that reactions of the following type were observed: $\gamma + N \rightarrow \pi + N$. In these reactions, the energy of the photons are directly absorbed by the nucleon and converted into the rest mass and kinetic energy of the meson.

As the bombarding energies of the nucleons and photons were increased, it was discovered that two or three or more $\pi$ mesons could be released from a single nucleon to produce reactions of the form $N + N \rightarrow N + N + \pi + \pi + \ldots$ or $\gamma + N \rightarrow N + \pi + \pi + \ldots$.

The number of mesons produced seems to be limited only by the amount of kinetic energy available that may be transformed into the rest mass energy of the mesons. The probability for a particular multiple meson production reaction decreases as the number of mesons increase, nevertheless reactions in which a large number of mesons are produced are still observed. The number of nucleons in production reactions always remained the same, however. But it is possible to have a reaction in which a nucleon-antinucleon pair is created such as $p + p \rightarrow p + p + \bar{p} + p$. In this reaction the number of nucleons remains the same since an antinucleon is counted as negative nucleon. The conservation of the number of nucleons in production reactions leads to the concept of assigning a quantum number to nucleons and pions called the baryon number. The proton and neutron each have baryon number 1, the antiproton and antineutron each have baryon number $-1$, and the pions and leptons have baryon number zero. The number of baryons is always conserved in all reactions including the decay of the neutron,

$$n \rightarrow p + e^{-} + \bar{\nu}.$$

The observation of meson production in nucleon-nucleon and photon-nucleon reactions more or less confirmed Yukawa's hypothesis that the strong interaction arises from the exchange of virtual mesons. This hypothesis received additional support by consideration of the electromagnetic properties of the proton and the neutron, namely their magnetic moment and charge distributions. The proton and the neutron are spin one-half particles like the electron and hence their electromagnetic properties have been described by Dirac's relativistic quantum mechanics. According to Dirac's theory, the magnetic moment of a spin half particle is related to its mass and charge such that the magnetic moment $\mu_e$ of the electron is given by the expression, $\mu_e = eh/2\pi m_e c$, where e is the charge and $m_e$ the mass of the electron. This expression provides a fairly accurate description of the electron's magnetic moment. Physicists have been able to show that the discrepancy between Dirac's prediction and the experimental values of the electron's moment can be explained by the effects of the virtual photons hovering about the electron.

In the case of the two nucleons, Dirac's theory predicted that the magnetic moment of the proton and the neutron, $\mu_p$ and $\mu_n$, respectively, are given by the following expressions, $\mu_p = eh/2\pi m_p c = \mu_N$ and $\mu_n = 0$. The experimental values of these quantities are in fact: $\mu_p = 2.79\ \mu_N$ and $\mu_n = -1.91\ \mu_N$. These discrepancies are quite considerable but can be understood in terms of the virtual meson hovering about the proton. The proton sometimes finds itself as a neutron and positive $\pi$ meson. This produces a current of positive charge which in turn generates a positive magnetic moment which, when added to the Dirac magnetic moment, gives a total value of 2.79 $\mu_N$. The neutron, on the other hand, finds itself at times as a proton and a negative $\pi$ meson. This produces a current of negative charge, which in turn generates the negative magnetic moment of the neutron.

In addition to qualitatively explaining the magnetic moments of the nucleons, Yukawa's picture of virtual meson exchanges also helps to explain the charge distribution of the proton and neutron. The charge distribution of the proton and the neutron is obtained by scattering electrons off of protons and neutrons. The electrons are assumed to be point particles without any structure. This is a prediction of Dirac's theory. Since the magnetic moment of the electron does not differ very greatly from that of the Dirac theory, we feel justified in assuming that the electron is a point of charge. This hypothesis can be tested eventually once an experiment can be devised in which we can scatter electrons off each other. In the meantime, we proceed with the assumption that any structure that the electron might have is small compared with that of the proton and neutron and hence can be ignored.

The Dirac theory also predicts that the proton will be a point charge and that the neutron will have absolutely no charge distribution at all. We expect major violations of this prediction, however, because of the presence of virtual mesons, which also carry charge. Probing the proton and the neutron with electrons is equivalent to taking a picture of the nucleons. Since we can not see the proton or the neutron with our eyes, we have devised a method of feeling our way about the nucleons by bouncing electrons off these particles and determining their shape by studying the distribution of angles into which the electrons are scattered as a function of the energy of the incoming electrons. These studies reveal the charge distribution for the nucleons.  The majority of the proton's charge is located at its core; however, the charge distribution extends to a distance of the order of $10^{-13}$ cm where it tails off. The edges of the proton's charge distribution may be interpreted as arising from the charge of the positive $\pi$ mesons hovering about the proton at a distance of approximately $10^{-13}$ cm. The charge distribution of the neutron, on the other hand, reflects the fact that the neutron spends part of its time as a proton and a negative $\pi$ meson. This explains the positive core of charge of the neutron, which is surrounded by a tail of negative charge that extends for about $10^{-13}$ cm about the neutron.

Yukawa's hypothesis of virtual meson exchange between nucleons ties together a great number of phenomena into a neat and tidy package by providing all at once an explanation of the range of the nuclear force, the production of mesons in nucleon collisions, the magnetic moments of the nucleons and their deviations from the Dirac theory, and finally the charge distribution of the proton and the neutron. Yukawa's theory, however, only provides a qualitative understanding of the strong interaction. It does not provide a quantitative description of the nuclear force between nucleons, nor does it indicate how many mesons are exchanged when nucleons interact. The theory of strong interaction is further complicated, as we shall shortly see, by the existence of other mesons in addition to the pion. These mesons will also be exchanged between nucleons and hence one has the additional problem of deciding which mesons are exchanged in a particular reaction. Finally, with the discovery of the pion and other mesons, and understanding of the nuclear force must include a study of the meson-nucleon and the meson-meson interactions as well as the nucleon-nucleon interaction. As is so often the case, the partial solution of one problem introduces a host of other problems as well.

The $\pi$ meson is a fairly long-lived particle on the scale of nuclear times. A nuclear reaction takes place in $10^{-23}$ seconds; therefore, the $10^{-16}$ sec lifetime of the neutral pion, $\pi^0$ and the $2.5 \times 10^{-8}$ sec lifetimes of the charged pions, $\pi^+$ and $\pi^-$ indicate that the pions are fairly stable particles. The $\pi^0$ decays via the electromagnetic interaction into two photons, $\pi^0 \rightarrow \gamma + \gamma$, whereas the charged pions decay via the weak interaction into muons and neutrinos: $\pi^+ \rightarrow \mu^+ + \nu_\mu$ and $\pi^- \rightarrow \mu^- + \bar{\nu}_\mu$. The charged pions can also decay into an electron and a neutrino or a muon, a neutrino and a photon. These two decay modes are much less frequent, each occurring only once in ten thousand decays. By comparing the rate of the reaction $p + p \rightarrow \pi^+ + d$ with the reverse reaction $\pi^+ + d \rightarrow p + p$, it was found that the spin of the pion is zero. Careful determinations of the masses revealed that the charged pions have identical masses of 139.6 MeV, but that the mass of the $\pi^0$ is 4.5 MeV less than the mass of the charged pions. The extra mass of the charged particles is due to the positive electric self-energy due to the potential energy of the meson's charge with itself. The differences in the lifetimes and masses of the different pions are due to these differences in charge. As far as the nuclear force is concerned, these particles are completely symmetrical as is the case with the neutron and the proton.

Because of the plentiful supply of pions produced when protons collide with other protons or nuclei, it was decided to use the pion as a probe to further explore the nature of the nucleus and the strong interaction. The very short lifetime of the pion, $2.5 \times 10^{-8}$ sec might appear to rule out making a beam of pions. Pions, however, are created with very high velocities. Even a pion with only 0.7 MeV kinetic energy has a velocity of 0.1 c and hence can travel on the average 0.1 c $\times$ 2.5 $\times$ $10^{-8}$ sec or 75 cm before decaying. This gives the experimentalists enough room to magnetically collect a beam of charged pions produced in the collision of a proton with a heavy target such as lead, and then magnetically direct that beam at a target such as liquid hydrogen. This enables the experimentalists to study pion-nucleon collisions. The experimentalists must use charged mesons, however, because they are unable to control neutral mesons with magnets. The experimentalist uses the charge on the $\pi^+$ and $\pi^-$ like a handle, leading them around with magnets.

The search for the intermediate meson predicted by Yukawa led to the discovery of a number of other particles, which like the pion and the nucleon also exert nuclear forces. These strongly interacting particles came to be known as strange particles for reasons, which will be clear shortly. The same type of cosmic ray studies which lead to the discovery of the muon and the pion lead to the discovery of four types of strange particles: the K meson or kaon, and two baryons, the $\Lambda$ and $\Sigma$ particles. The $\equiv$, or cascade particle and the $\Omega$ particle, were discovered using accelerators.

The K mesons or kaons all have a mass of approximately 500 MeV, which is about 3.5 times, the mass of the pion. There is a charged kaon, the $K^+$, and a neutral kaon, the $K^\circ$. In addition these two particles have antiparticles, the $K^-$ and the antiK$^\circ$. The $K^\circ$ and the antiK$^\circ$ are two separate particles, which behave differently. The K mesons, like the pions, have baryon number zero. The K mesons are different than the $\pi$ mesons, however, in that the $\pi^-$ is the antiparticle of the $\pi^+$ but the $\pi^\circ$ is its own antiparticle. There are, therefore, only three pions but four kaons, the $K^+$ and the $K^0$ and their antiparticles. The $\Lambda$, with mass 1115 MeV; the $\Sigma$ with mass 1190 MeV; the $\equiv$ with mass 1315 and the $\Omega$ with mass 1670 all have masses greater than the proton and the neutron whose masses are approximately 940 MeV. These particles are referred to as hyperons because they have masses greater than the nucleons. Because they all decay eventually into one nucleon, they are baryons each with baryon number equal to 1.

The reason that the strange particles received their nickname is because of the fact that they must be produced in pairs and that they carry a strange charge called strangeness. They are produced quite readily by the strong interaction as long as strangeness is conserved, however, they decay very slowly via the weak interaction into pions or muons or both if they are kaons and into nucleons and pions if they are hyperons. We shall return to the story of their decay when we discuss the weak interaction. For the purposes of the present discussion it is important to note that the decay of all the strange particles is of the order of $10^{-10}$ or $10^{-8}$ sec that is characteristic of the weak interaction and that these decays violate strangeness conservation that was characteristic of their production via the strong interaction.

An explanation of why the strange particles are produced strongly yet decay weakly was first made by A. Pais who suggested that these particles could only be produced strongly in pairs and since they decay singly into non-strange particles, they can not decay strongly but only weakly. Gell-Mann and Nishijima each separately worked out an identical scheme in which they assigned a strangeness quantum number to each elementary particle. The pion and nucleon which do not behave in a strange way were assigned 0 strangeness. The nucleon has two charge states +e and 0 and hence has a center of charge at e/2. The $\Sigma$ particle has three charge states, +e, 0 and −e and hence has a center of charge at 0, as does the $\Lambda$ particle which has only one charge state, 0. The center of charge of the $\equiv$ is at −e/2 since it has two charge states, $\equiv^-$ and $\equiv^0$. The $\Omega^-$ has only the charge state −e and hence the center of charge at −e. these particles were assigned positive strangeness according to how many units of e/2 their charge center differed from that of the nucleon.

In a similar manner the strangeness of the kaon was determined by the fact that its center of charge +e/2 for the $K^+$ and $K^0$ is one unit from the 0 center of charge of the $\pi^+$, $\pi^0$ and $\pi^-$ and hence has strangeness −1. The strangeness of the $K^-$ and anti$K^0$ on the other hand is +1 since their center of charge is at −e/2. According to the Gell Mann–Nishijima scheme the strong interaction conserves strangeness and baryon number and the weak interaction violates strangeness conservation but still maintains baryon conservation. All the decays of the strange particles are weak because there are no energetically possible decay modes, which do not violate strangeness. The $\Lambda$ does not have enough energy to decay into a $K^-$ and a proton. All of the decay modes of the strange particles involve products with one less unit of strangeness. Nishijima and Gell Mann therefore predicted that the strong production of strange particles would always involve a combination of particles such that the total strangeness before and after the collision was the same. Hence the $\Lambda$ or $\Sigma$ would be produced with a $K^+$ or a $K^0$ such as $\pi^- + p \rightarrow K^0 + \Lambda$ or $\pi^- + p \rightarrow K^+ + \Sigma^-$. But a reaction such as $\pi^- + p \rightarrow K^- + \Sigma^+$ could never take place, or for that matter $p + p \rightarrow p + n + K^+$. Their predictions were confirmed almost immediately by the results from the 3 GeV Cosmotron built at Brookhaven Labs.

The first baryons that were discovered, namely, p, n, $\Lambda$, $\Sigma^+$, $\Sigma^0$, $\Sigma^-$, $\Xi^-$, and $\Xi^0$ all were spin ½ particles with baryon number 1. The first mesons that were discovered, namely, $\pi^+$, $\pi^0$, $\pi^-$, $\eta$, $K^+$, $K^0$, $K^-$, and antiK$^0$ were all spin 0 particles with baryon number 0. The η meson was an unstable meson that decayed via the electromagnetic interaction into 3 π mesons or a π meson and two photons. The baryons and mesons, which form the class of hadrons or strongly interacting particles can be grouped in families in which there are 8 members, 3 of which have the same mass but different charges, two sets of two with the same mass but different charges and one member with its own unique mass. As higher energy collisions of protons with protons and mesons with protons were performed new sets of baryons and mesons with the same structures but greater spins began to be discovered. Among the mesons there were spin 1, 2, ... sets and among the baryons there were spin 3/2, 5/2, ... most of which came in families of 8. Among the baryons there was another family of 10. Murray Gell Mann and Yuval Ne'eman independently showed that all of these families possessed the same symmetry, namely that of SU(3), the special unitary group of degree 3.

## Quarks

Murray Gell Mann and George Zweig independently in 1964 showed that this symmetry could be explained if the hadrons were made up of more elementary particles, which Gell Mann called quarks. In his book, *The Quark and the Jaguar*, he describes how he came up with this term, "In 1963, when I assigned the name 'quark' to the fundamental constituents of the nucleon, I had the sound first, without the spelling, which could have been 'kwork'. Then, in one of my occasional perusals of *Finnegans Wake*, by James Joyce, I came across the word 'quark' in the phrase 'Three quarks for Muster Mark'."

Consistent with the SU(3) symmetry there are 3 kinds of quarks: up down and strange of which the first generation of hadrons are composed. The quarks were spin ½ particles that had the charges respectively of +2/3e, −1/3e and −1/3 e. They differ from hadrons and leptons such as electrons and muons in that their charges come in multiples of e/3 instead of ±e as is the case for all other elementary particles. The baryon number of quarks and antiquarks is 1/3 and −1/3 respectively whereas for all other elementary particles their baryon numbers are +1, 0 or −1. There was also a set of 3 antiquarks with the opposite charges of the quarks. The strange quark has strangeness number +1 and the antistrange quark has strangeness number −1. Baryons are made up of 3 quarks, antibaryons of 3 antiquarks and mesons of a quark-antiquark pairs. The proton is made up of 2 up quarks and one down quark so its charge is $+2/3$ e $+2/3$ e $-1/3$ e or $+e$ and the neutron is made up of one up and two down quarks so its charge is $+2/3$ e $-1/3$ e $-1/3$ e $= 0$. The $\Sigma^+$ is made up of 2 up quarks and one strange quark so its charge is $+2/3$ e $+2/3$ e $-1/3$ e or $+e$ and its strangeness number is $+1$. The $\pi^+$ is made up of an up quark and o ne antidown quark so its charge is $+2/3$ e $+1/3$ e or $+e$. The K+ is made up of an up quark and one antistrange quark so its charge is $+2/3$ e $+1/3$ e or $+e$ and its strangeness number is $-1$. The K⁻ is made up of an antiup quark and one strange quark so its charge is $-2/3$ e $-1/3$ e or $-e$ and its strangeness number is $+1$. All of the hadrons can be accounted for in terms of quarks by making combinations similar to these.

In addition to the families of 8, SU(3) also predicted that there would be baryon families of ten particles consisting of the following combinations of the up (u), down (d) and strange (s) quarks: uuu, uud, udd, ddd, uus, uds, dds, uss, dss, and sss. The sss combination corresponds to the $\Omega^-$ and the uuu to the spin 3/2 $N^{*++}$.

As scattering experiments were conducted at still higher energies physicists encountered hadrons that were composed of other kinds of quarks so that in addition to the first three quarks used to describe the first generation of hadrons three more types of quarks were encountered, named charm, bottom and top as well as their associated antiquarks. The charm quark and the top quark each have a charge of $+2/3$ e while the bottom quark has a charge of $-1/3$ e.

Although one can explain the SU(3) symmetry of hadrons in terms of quarks, a quark has never been seen in isolation because of the phenomenon of confinement, which is not totally understood. It is just an empirical fact that quarks never can be separated from the hadrons in which they exist. The closest physicists have come to seeing a quark is when they are produced in high energy collisions and one see three jets of many hadrons clustered together. The force between quarks that gives rise to their confinement is due to the exchange of gluons. Gluons play the same role in the quark-quark force that the photon plays in the force between charged particles and mesons play in the force between nucleons. Like photons gluons have spin 1 but they come in eight different colour charges. The colour in colour charge is not the actual colours of visible light but they are a metaphorical way of describing the different states of gluons. Quarks also have three different colour charges and antiquarks have three kinds of anticolours.

The interactions of quarks and gluons are described by Quantum Chromodynamics (QCD), which is a theory of the fundamental force of nature, the strong interaction of baryons and mesons, i.e. hadrons, and is patterned on Quantum Electrodynamics. It is an essential part of the Standard Model of particle physics and describes and explains a huge body of experimental data collected over many years.

## The Weak Interaction

The weak interaction is one of the four basic forces of nature along with the strong interaction, gravity and electromagnetism. It is the force that accounts for the decay of the strongly interacting hadrons and is extremely weak with a strength of $10^{-11}$ that of the electromagnetic force and $10^{-13}$ of the strong interaction. The first example of the weak interaction that was encountered was the beta decay of a free neutron into a proton, an electron and an antineutrino, $n \rightarrow p + e + \underline{v_e}$. [A note on notation: the neutrino will be represented by $v_e$ and the antineutrino by $\underline{v_e}$] The neutrino and antineutrino are uncharged elementary particles with a minuscule, but nonzero mass that travels very close to the speed of light and is difficult to observe because they pass through ordinary matter for the most part without interacting. The existence of this particle was first suggested by Wolfgang Pauli to explain the lack of conservation of energy, momentum and angular momentum when neutron decay into a proton and an electron was first observed. Neutrino was first detected in 1956 when they were observed in induced beta decay ($\underline{v_e} + p \rightarrow n + e^+$) in which antineutrinos colliding with protons produced pairs of neutrons and positrons that could be detected.

Another example of an early encounter of the weak interaction was the decay of the muon into electrons and neutrinos and the decay of the pions into muons and e-neutrinos, $v_e$ and $\mu$-neutrinos, $v_\mu$. Muons were first discovered in cosmic ray showers and were observed to decay with a mean life of $2.2 \times 10^{-6}$ seconds as follows: $\mu^- \rightarrow e^- + \underline{v_e} + v_\mu$ and $\mu^+ \rightarrow e^+ + v_e + \underline{v_\mu}$. Pions were also first discovered in cosmic ray showers and were observed to decay with a mean life of $2.6 \times 10^{-8}$ seconds as follows: $\pi^+ \rightarrow \mu^+ + v_\mu$ and $\pi^- \rightarrow \mu^- + \underline{v_\mu}$.

The electrons and the muons and their associated neutrinos form a family of particles known as leptons. The muon is basically a heavy electron with a rest mass energy of 105.7 MeV compared to the 0.5 MeV rest mass energy of the electron. There is one more member of this family that has so far been discovered namely the tauon and its associated tau-neutrino. The rest mass of the tauon is 1,777 MeV and it has a mean lifetime of $2.9 \times 10^{-13}$ seconds and decays via the weak interaction into leptons and sometimes into hadrons.

The weak interaction is mediated by the exchange of three kinds of bosons the $W^-$, the $W^+$, the antiparticle of the $W^-$ and the Z, which is its own antiparticle because it has no charge. All three bosons are very short-lived with a mean lifetime of about $3 \times 10^{-25}$ seconds. They are extremely heavy particles with masses almost 100 times that of the proton (940 MeV), namely with a rest mass energy of 80,400 MeV for the W and 91,200 MeV for the Z.

### Summing up the Interactions of Elementary Particles

There are four basic interactions in the universe, gravity, weak, electromagnetic and strong. The $W^+$, $W^-$, and Z bosons play the same role for the weak interaction as the photon does for the electromagnetic force and together they form the four gauge bosons of the electroweak interaction and hence unite the electromagnetic force and the weak interaction as one force. The electroweak interaction and gravity are forces that impact all the particles of the universe as opposed to the strong interaction, which only operates on hadrons consisting of baryons and mesons.

Given that hadrons are composites of quarks and antiquarks the number of known elementary particles can be reduced to the following sets of quarks, leptons and bosons, namely

1. The following quarks and their associated antiquarks: up, down, strange, charm, top and bottom.
2. The following leptons and their associated antiparticles: e, $\mu$, $\tau$, $v_e$, $v_\mu$, and $v_\tau$.
3. The four gauge bosons, $W^-$, $W^+$, Z and $\gamma$.

Some elementary particle physicists believe that there is another boson out there called the Higgs particle, whose existence would explain the difference in mass of the four gauge bosons. Only the new ultra-high energy accelerators will be able to find it if it indeed exists.

There are also attempts being made now to unite the electroweak interaction with quantum chromodynamics or the strong interaction. There is also the work being done in string theory to unite all four forces. String theory is a heroic attempt to create a unified field theory in which the gravitational force described by the General Theory of Relativity could be united with the three basic forces described by quantum chromodynamics and the electroweak interaction. The basic idea is that the elementary particles are made up of vibrating one dimensional strings or membranes embedded in a multidimensional space, which in some theories has 11 dimensions. The theory has been pursued for over 30 years and has not been able to make one prediction capable of experimental testing. Not only that — there is not a unique string theory but one can with this approach generate $10^{500}$ solutions to the theory.

In my opinion and that of a number of other physicists much more prestigious than I, string theory is not a science because it cannot make any predictions and hence cannot be falsified. For readers who would like to read a deeper critique of string theory I would recommend my friend Lee Smolin's (2006) book, *The Trouble with Physics*. For those readers who have read about string theory and might be disappointed by my rather short treatment of this topic I apologize but I cannot write at length about an approach I do not believe in. It would be great if string theory were to work but one has to be realistic. When I conducted my research in 1965 that led to my PhD at MIT I performed an analysis of the reaction $\pi^- + p \rightarrow \pi^0 + n$ to show the existence of Regge poles, singularities in the complex angular momentum plane. Regge poles provided an excellent description of high energy scattering but in the end they did not advance our fundamental understanding of the nuclear force. They were not wrong; they just were not very useful as the following excerpt from the Wikipedia article on Regge theory indicates: "As a fundamental theory of strong interactions at high energies, Regge theory enjoyed a period of interest in the 1960s, but it was largely succeeded by quantum chromodynamics." C'est la vie for many ideas in physics! This is the nature of physics, not every good idea in physics turns out to be useful.

# Chapter 25

# Cosmology and the Universe: The Big Bang, Dark Matter and Dark Energy

Up to this point we have concerned ourselves with the general principals and concepts of physics focusing both on their historic development and their general impact on society. We have not explored the physical universe, which constitutes our cosmic environment. We have spent some time understanding the microscopic structure of matter but we have not yet investigated macroscopic structures such as the continents, the oceans, the Earth, the planets, the stars, the Milky Way Galaxy or the universe as a whole. In this section of the book we shall tour the cosmos. We shall investigate the nature and the origin of the planet Earth, the heavenly bodies that surround it and the universe itself.

## Images of the Universe

We begin by examining the structure of the universe. It is interesting to follow the historical development of man's concept of the universe. We have already looked at some of the earlier notions of the cosmos. The story of Kujum Chantu as told in Chapter 2 represents one of the earliest geocentric pictures of the universe. The universe in these primitive models consists of a flat Earth above which lies the vault of the heavens frequently described as the inside of a giant cosmic bowl. Below the surface of the Earth lies the dark underground region, the domain of the dead. As more sophisticated concepts of the universe developed during the Greek era the Earth was considered to be a sphere about which the heavenly bodies — the Sun, the Moon, the five planets and the stars — revolved. At this stage of development the universe did not extend much beyond the dimensions of the solar system, the distance to the stars and the planets being regarded as basically equivalent.

With the Copernican revolution the Sun became the center of the universe about which the Earth and the other planets orbited. The stars were thought to be fixed objects, which surround the Sun on all sides. The dimensions of the universe greatly increased with the adoption of the Copernican picture. Since the Earth is no longer at rest, one should observe the aberration of starlight (see Chapter 13). Copernicus was unable to observe this phenomenon without a telescope, but concluded that the distance to the stars was far greater than anyone had previously thought.

Copernicus and his contemporaries still had no idea of the true distances to the stars. With the use of the telescope the distances to the nearest stars were determined and our closest neighbor was found to be four light years away. A light year is the distance a particle of light travels in one year and corresponds roughly to $10^{16}$ meters or 9 trillion kilometers. The dwarf planet Pluto is 12 light hours from the Sun while the Earth is only 8 light minutes away. The nearest star is 10,000 times as far away from the Sun as the Earth.

The telescope revealed that the sky was filled with stars most of which cannot be seen with the naked eye. The Milky Way when gazed at through a telescope was seen to consist of millions and millions of stars so close to each other in the sky that they appear as a cloud. By counting the number of stars in the various parts of the sky astronomers discovered that the stars are not distributed homogeneously in the heavens. Instead they found that the stars are distributed in an enormous disk-like structure with a bulge in its center. When looking at the Milky Way we are looking at the side of the disk. Because we are looking into the plane of the disk we see more starts than we see in any other direction. This disk-like structure of stars is called a galaxy after the Milky Way since the Greek word for milk is "galaxias".

The Milky Way Galaxy at first was believed to compose the entire universe. The dimensions of the galaxy-universe were explored and by 1920 the radius of the galactic disk was determined and found to be about 50,000 light years. The Sun was shown to be near the edge of the disk some 26,000 light years from the galactic center. It was also discovered that the Sun along with other stars orbits the center of the galaxy under the gravitational influence of the galactic nucleus, which contains the majority of the mass of the galaxy. The period of the Sun's orbit about the galaxy is 240 million years. With the observation of the galactic universe the perceived dimensions of the universe greatly increased. The Sun was dethroned from its special position at the center of the universe. It became one of $10^{11}$ stars occupying some obscure corner of the galaxy-universe. The new universe from an aesthetic point of view was not pleasing because of its lack of uniformity and homogeneity.

This notion of the universe was not destined to have a long life, however. Just as the nature of the Milky Way Galaxy was being understood, Edwin Hubble began his study of nebulae, the fuzzy cloud-like non-stellar objects found in various parts of the sky. He found that not all of these fuzzy patches of light are the same type of objects. Many are luminous clouds of gas several light years across lying within our galaxy. Other nebulae lie outside our galaxy and are galaxies in their own right. Their distance from us varies from 1 million light years to 13.7 billion light years, the edge of the observable universe. These distant galaxies also contain millions and millions of stars gravitationally bound to each other and orbiting their galactic nuclei.

It is estimated that the observable universe contains approximately $10^{11}$ galaxies, which is as many as the number of stars contained in a single galaxy. The observable universe, therefore, contains $10^{22}$ stars altogether. The galaxies are found to form clusters and superclusters in which the galaxies are gravitationally bound to each other. The number of galaxies per cluster varies from 20 in our own local group to 10,000 found in a cluster 1.23 billion light years away in the constellation Como Berenices. There is some evidence that clusters of galaxies form gravitationally bound super clusters. For the purposes of our discussion, however, we will take the clusters as the largest sub units of the universe.

Galaxies and their clusters are distributed homogeneously throughout space suggesting that our universe is symmetrical and uniform after all. Perhaps the most interesting feature of the clusters is the fact that every single observed cluster is moving away from our local group or cluster with a velocity proportional to its distance from us. The universe seems to be expanding with us at the center. Because of the proportionality of the distance to a cluster and its recessional velocity, the universe viewed from any other cluster will appear the same as it does to us. It too will be the center of an expanding universe. All the clusters will be receding from it with velocities proportional to their distances. The motion of the clusters in three dimensions may be likened to the two dimensional motion of the dots on the surface of a balloon being blown up. From the point of view of any dot all the other dots are moving away from it with a velocity proportional to their distance. An analogy in three dimensions often cited to describe the expanding universe is the one in which the galactic clusters are likened to the raisins in a raisin bread which is rising.

An observer sitting in any cluster of galaxies will observe all the other clusters receding away from him at exactly the same rate as his counterpart on any other cluster. The universe appears identical from the point of view of any cluster. The universe is, therefore, more than just symmetrical; it is also isotropic (the same in all directions) and uniform. One part of the universe is the same as any other part. This assertion about the nature of the universe is known as the cosmological principle and forms the basis of all contemporary descriptions of the universe.

This more or less completes our description of the universe, as we presently know it. It is a universe, which displays a hierarchical structure. Stars and their satellites, i.e. planets are embedded in galaxies, which in turn are embedded in clusters. One of the elements we left out of our description of the universe is the immense space between the planets in the solar system, between the stars within a galaxy, between the galaxies within a cluster and between the clusters within the universe. Venus, our closest neighbor in the solar system, is never less than 40 million kilometers away. The nearest star is 4 million light years away; the nearest galaxy is 1 million light years away. The universe is, therefore, basically space but not empty space.

The space between the stars and the galaxies is filled with a very dilute gas. The density of the interstellar gas is estimated to be $10^{-24}$ grams per cubic centimeter or one atom per cubic centimeter. The density of water is one gram per cubic centimeter or $10^{24}$ atoms per cubic centimeter. The interstellar gas is primarily hydrogen and helium, which represent about 68% and 30% of the gas by weight, respectively. The remainder of the gas consists of sodium, potassium, calcium, carbon, oxygen, nitrogen, iron and titanium. Even though the density of the interstellar gas is very low, there is a great deal of it because of the large volume of the galaxy, which is approximately $10^{67}$ cubic centimeters. The total amount of gas is, therefore, $10^{-24} \times 10^{67}$ grams or $10^{43}$ grams which represents $10^{10}$ solar masses or 10% of the total mass of the galaxy. In addition to the interstellar gas there are also interstellar dust particles whose total mass and density is approximately one percent that of the interstellar gas. It is the interstellar gas and dust, which provides the material for the formation of new stars within the galaxy. The space between the galaxies also contains gas and dust particles with densities believed to be about one millionth of their interstellar counterparts within a galaxy.

Interstellar and intergalactic space is filled with dust and gas and with isotropic fluxes of low energy photons, neutrinos and cosmic rays, which consist primarily of high-energy protons and electrons. These elementary particles do not interact very strongly with the interstellar dust or the stars and play little or no role in the formation of stars or galaxies. They do provide information, however, regarding the origins of the universe. In addition to all of these observable forms of energy and matter we will discover that there is also two mysterious substances in the universe that we will learn more about later in this chapter. They are dark matter and dark energy.

## The Expanding Universe

The vastness and the complexity of the universe is a source of mystery. One wonders how the universe came into existence and how structures such as clusters, galaxies, stars and planets were formed. Perhaps the most mysterious feature of the universe to be explained is its expansion, which recent observations seem to indicate is actually accelerating. Before attempting this awesome task, let us first understand how the expansion of the universe was determined phenomenologically. This involves understanding how the distances to stars, galaxies and clusters are determined and how the velocities of these objects are measured.

The determination of the velocities of astronomical bodies in the direction along the line of sight of the object either towards or away from Earth is quite simple. The motion transverse to the line of sight is more difficult to determine especially as the distance to the object increases. Stars and interstellar gas are composed of exactly the same atoms found on Earth, such as hydrogen and helium. The frequency of light emitted by atoms is, therefore, known. If a star is moving away from or towards us, the frequencies it emits will be Doppler shifted. This is similar to the Doppler shift of a train whistle. Just as the frequency of the train whistle increases as the train enters the station and decreases as it leaves, the frequency of the star's electromagnetic radiation decreases or increases depending on whether it is moving away from or toward the Earth.

For the stars of our own galaxy we see light shifted both up and down. Some stars are moving away from us in which case the frequency of their emission lines shift toward the red while some are moving toward us and hence have emission lines shifted towards the blue. For the stars of the other galaxies in our own local clusters both red and blue shifts are also observed. For the stars of other clusters, however, only red shifts are observed indicating that these stars are all moving away from us.

The determination of the velocity of a star in the direction perpendicular to the line of sight is made by measuring the change of the angle of observation with time and knowing the distance to the star. The transverse velocity can only be measured for nearby stars because of the limitations of making accurate measurements of the angle of observation.

The methods of measuring distances differ for each scale of distance measured. The methods of measuring larger distances frequently depend on the methods of measuring smaller distances and, therefore, we begin our survey with measurement of distance on the scale of the solar system. The distance between Earth and Mars was determined using the technique of triangulation. The position of Mars was observed from two points on Earth whose distance from each other was known. By observing the angle at which Mars appears in the sky with respect to the distant fixed stars at the two points on Earth, one is able to determine the triangle made by the two points on Earth and the planet Mars. One can then determine the distance to Mars from this triangle. Once this distance is known one can then use Kepler's laws of planetary motion to determine the other distances within the solar system including the distance of the Earth from the Sun.

With the knowledge of the diameter of the Earth's orbit about the Sun, one can use the triangulation method to measure the distance of nearby stars. One observes a star from the same point on Earth at 6-month intervals. The two positions of the Earth at this six-month interval determine the base of the triangle. Measurements of the angle of observation of the star at each of these positions enable one to complete the triangle and determine the distance to the star. This method works for stars up to 150 light years away. For distances greater than this, the angles of observation cannot be measured accurately enough to make a proper determination of the stellar distance.

For distances greater than 150 light years one must make use of Cepheid variables, a class of stars whose luminosity varies periodically with time. The observation of a number of Cepheid variables in the Large Magellanic Cloud, a nearby galaxy, revealed a relation between the absolute luminosity of this type of star and the period of the variation of its luminosity. The observed luminosity of a star is equal to its absolute luminosity divided by the square of the distance from the observer. By calibrating the Cepheid variables in our own galaxy whose distance could be determined, the distance to galaxies containing Cepheid variables could be determined using the relation between absolute and observed luminosities.

For those galaxies without a Cepheid variable another technique for determining distances was developed. By observing many galaxies whose distances are known one is able to determine the absolute magnitude of the brightest star of each of these galaxies. The brightness of these stars, the brightest of their galaxies, is more or less the same. One can, therefore, estimate the distance to a galaxy with no Cepheid variable by determining the observed brightness of the galaxy's brightest stars and assuming its absolute brightness is the same as that of the other brightest stars. This same type of comparison is also used to estimate the distance to a cluster of galaxies. In this case one measures the luminosity of the brightest galaxy of the cluster and assumes the absolute luminosity is the same as that of the brightest galaxies of other clusters whose distances are known.

Hubble discovered that for distances greater than 30 million light years that the recessional velocity of a cluster of galaxies is proportional to its distance from us. He found that the velocity of the cluster increases 30 kilometers per second for every million light years away the cluster is found. A cluster 100 million light years away, for example, will have a velocity equal to 30 kilometers per second times 100 which is 3000 kilometers per second or 1/100 the velocity of light. Most clusters are observable up to distances of 3 billion light years after which they become too faint to be observed. Their velocities at these distances is approximately 90,000 kilometers per second or 3/10 the velocity of light. There is a special class of galaxies known as exploding galaxies which are much brighter than normal galaxies and hence, can be seen further away. These galaxies have been observed as far away as 6 billion light years with receding velocities as high as 6/10 the velocity of light. At first the universe was thought to be expanding at a constant rate but recent observations seem to indicate that the rate of expansion is increasing or accelerating.

## Cosmological Speculations

What lies beyond these galaxies that are just barely visible? It is impossible to observe galaxies further away because they are not bright enough to be seen but there is no reason to believe that galaxies suddenly stop existing at 6 billion light years. How far does the universe extend, however? Is the universe finite in extent as the Greeks imagined or is it infinite as Bruno and the deists believed? One may also ask whether the universe will continue to expand forever and if so into what? Or will it contract eventually and then re-expand as some claim. And what is causing the acceleration of the expansion? These questions are intriguing but at the moment are matters of scientific speculation. Various models of the universe provide different answers to these questions. One of the more popular approaches introduces the notion of dark matter and dark energy, which we will say more about presently.

The expansion of the universe provides some clues about the size of the universe depending on how one regards the origin and nature of this expansion. If the velocity of the clusters continues to increase with distances beyond the 6 billion light year boundary for observing exploding galaxies, then at a distance of 13.7 billion light years the velocities of the clusters will equal the velocity of light. Light from an object, no matter how bright it is, could never reach us if its velocity equals the speed of light because of the Doppler shift. The frequency of the light emitted by such an object would be shifted all the way to zero. The observable universe, therefore, would have a finite radius of 13.7 billion light years.

What lies beyond this radius of 13.7 billion light years? Are there stars and galaxies, which we will never be able to detect because their light is red shifted to zero? Or does the universe end abruptly at this point with nothing lying beyond the edge of the 13.7 billion light year radius but empty space? Or does the universe close back upon itself so that a space traveler could never come to the edge of the universe? In such a closed universe if I traveled far enough I would return to the point of my origin. This is exactly what would happen if I were to walk far enough along the surface of the Earth. Or do I come to the edge of the universe and fall off as the sailors who first crossed the Atlantic once feared?

These questions are part of the domain of cosmology, the branch of astronomy, which tries to understand the universe as a whole. Cosmologists do not have hard and fast answers to these questions. They have developed theories each of which answers these questions differently. Almost all cosmologists believe that an explanation will involve some variation of the Big Bang theory proposed by Georges Lemaitre in 1927. The only alternative, the Continuous Creation or Steady State Theory, proposed by Bondi, Gold and Hoyle in 1948 has more or less been dismissed but it is worth mentioning for historic reasons. According to the latter theory the universe is in a steady state. It is infinite in extent. Despite its steady expansion, its density remains essentially constant because of the continuous creation of matter at just the rate required to produce an equilibrium state, which requires that one proton per cubic meter, is created every million years.

The Big Bang theory postulates a much more violent universe. The proponents of this theory believe that the universe began 13.7 billion years ago as a point of energy and it has been expanding ever since. If this is true, then projecting backwards in time the universe will shrink. Eventually one comes to a point in time when the universe collapses down to a point. If this is true, the collapse to a point is reached by going 13.7 billion years backwards in time, which the proponents of the Big Bang theory believe was the beginning of the universe. They believe that all the matter/energy of the universe was contained in this single point 13.7 billion years ago, which exploded because of its inherent instability. There are various explanations of how this singularity came into being.

The universe has been expanding ever since the Big Bang. Most of the expansion is due to the fact that chunks of matter were propelled from the central point with varying relative velocities. These chunks of matter eventually evolved into clusters. Those chunks of matter, which were ejected with velocities near the speed of light with respect to us, have since traveled to the edge of our observable universe 13.7 billion light years away. Since no matter can travel faster than the speed of light no matter lies beyond the edge of the observable universe. Ignoring the acceleration of the expansion the distance of a cluster from us is basically proportional to its velocity relative to us simply because the faster a chunk of matter was traveling just after the initial explosion the further it traveled. Because all velocities are relative, no one cluster may be considered to be the center of the universe about which all the other clusters expand. Rather every cluster may be considered the center of the universe.

The Big Bang theory provides a reasonable explanation of the universe as well as a number of other cosmological phenomena we shall shortly examine. The adoption of the theory still leaves open the question of the finiteness of the universe. Does the universe close in upon itself or is the region beyond the 13.7 billion light year radius empty space, which extends to infinity? Other questions also present themselves. Will the universe continue to expand forever eventually becoming infinite in size after an infinite time has passed? If this is true, what happened before the creation of the universe? Was there such a thing as time before the Big Bang?

Another possibility that the theory allows is that the expansion would eventually slow down as a result of the gravitational attraction of all the clusters for each other. It is even possible that the expansion could be reversed and that the universe would eventually shrink back into a point and explode again starting another cycle of the universe. If this is true which cycle are we currently living through? Does history in the cosmic sense repeat itself? This possibility seems remote as the expansion of the universe seems to be increasing rather than decreasing.

Still another possibility is that the universe began an infinite time ago as an infinitesimally dilute gas of infinite extent, which began collapsing. Then 13.7 billion years ago the universe had collapsed down to a point and exploded. The universe is now expanding and will continue to expand forever eventually returning to the state from which it began. In this model there is only one cycle of the universe.

This model and the model just discussed in which the universe is continually oscillating between expanding and shrinking phases, have one feature in common. In both models the universe always was and always will be. I personally find this an appealing principle because of the conservation of energy. If the universe were suddenly to start at time zero from nothing, I would want to understand how such a colossal violation of energy conservation could take place. If the total energy of the universe were decreasing, one could go forward in time to when there would be no energy and hence the end of the universe. If the total energy of the universe were increasing, then one could go backward in time to when there was no energy or the point when the universe just began. But since total energy is conserved I believe that the universe has neither a beginning nor an end. It always was and always will be "an ever-kindling fire" as Heraclitus described it.

There is one aspect of the observable universe that the Big Bang theory cannot take into account and that is the fact that the rate at which the universe is expanding is accelerating. The acceleration is believed to be due to a long-range repulsive force, which is being attributed to dark energy and dark matter. We will discuss these possibilities when we return to the cosmological implications of Einstein's General Relativity Theory and the role of the cosmological constant $\Lambda$.

## Cosmological Implications of General Relativity

Cosmology endeavors to understand the properties of the universe in space and time. Consideration of Einstein's General Theory of Relativity is essential for understanding why the expansion of the universe is accelerating and for determining if the universe is finite and closed or infinite and open. General relativity fuses space and time into a four dimensional continuum. The properties of the space-time continuum are related to the gravitational interaction of matter, which is the one interaction that determines the structure of the universe as a whole. As we will see the other three basic forces of the strong, electromagnetic and the weak interactions play an important role in the dynamics of the early universe.

Einstein showed that the gravitation interaction of matter warps, curves or bends the four-dimensional space-time continuum changing the very structure of space and time. Using this notion he successfully predicted the bending of starlight by the Sun and explained the advance of the perihelion of Mercury.

Einstein adopted his field equations to deal with the universe as a whole and found a solution to these equations corresponding to a static universe. A static universe is one whose size remains fixed. De Sitter found a second solution to Einstein's equation also corresponding to a static universe. Both these models were found to be in contradiction with observations, however, and had to be abandoned. At approximately the same time Hubble was demonstrating experimentally that the universe is expanding, the Russian mathematician A. Fuedman showed that solutions in which the universe was either expanding or shrinking were consistent with Einstein's field equations. From this result, as well as Hubble's experimental work, the notion of the expanding universe developed.

The question of the finiteness or infiniteness of the universe within the framework of general relativity depends on the nature of the curvature of the four dimensional space-time continuum. If the curvature of the space-time continuum is negative, then the universe is open and infinite. If the curvature is positive, then the universe curves back upon itself in the four-dimensional space and hence is closed and finite. On a clear day one could see the back of one's head if one stood still for 13.7 billion years. The analogy of a closed and finite space in two dimensions is a circle, which closes back upon itself to form a one-dimensional closed space. The analogy in three dimensions is the surface of a sphere, which closes back upon itself forming a two-dimensional closed space. If the universe curves back upon itself in the four dimensional space-time continuum, then it forms a closed three dimensional space, the surface of a four dimensional sphere. If the universe is closed, then a photon emitted in a given direction if unimpeded will eventually return to the point of its origin having traversed a complete hypercircle. This is exactly analogous to the situation on the surface of a three dimensional sphere. A creature crawling on the surface of the Earth will eventually return to the point of its departure having traversed a great circle because the two-dimensional surface of the Earth curves back on itself.

Einstein showed that the nature of the curvature of the four dimensional space-time continuum depends on the density of matter within the universe. If the density of the universe is small, then the curvature is negative but if the density is great, then the curvature is positive and the universe is finite and closed. It is difficult to determine the density of the universe because of our ignorance of intergalactic space. There is also the question of dark matter and dark energy needed to explain the acceleration of the rate of expansion, but more of that later. We can estimate the amount of galactic matter, however. If the space between the galaxies is void, then the density of the universe is too low to curve the space-time continuum positively and the universe is open and infinite. The universe will continue to expand forever because the forces of gravity will never be strong enough to pull it back together again.

If the intergalactic medium contains a sufficient amount of matter then it is possible that the curvature is positive. In order for the universe to close upon itself the amount of extra galactic material would have to be equal to 10 times the amount of matter in the galaxies. Because of the vast expanses of intergalactic space, the density of matter required to furnish this much mass is only $10^{-29}$ grams per cubic centimeter or one hydrogen atom for every 100 cubic centimeters. This is still a rather rarefied medium when one considers that the density of a galaxy is $10^{-24}$ grams per cubic centimeter. At the moment the density of intergalactic space is unknown and, hence, it is still an open question as to whether our universe is open and infinite or finite and closed. There is also the issue of dark matter and dark energy that must be examined.

## The Big Bang Theory

The General Theory of Relativity provides a general framework for describing the universe as a whole. It does not, however, constitute a complete description since the initial condition of the universe must be specified in order to solve the field equations. There are literally an infinite number of possibilities. Cosmologists attempt to hypothesize a set of initial conditions that will provide a solution to the field equations, which matches our astronomical observations. Each set of conditions defines a model or a theory, which must be tested. The Big Bang theory is one example of such a model, which we will soon discover, has many different variations. The motivation of the Big Bang theory is that it provides a simple explanation of the expansion of the universe, in particular, the approximate linearity of velocity and separation.

## Genesis

According to the Big Bang theory the universe or the present cycle of the universe began 13.7 billion years ago as a singularity or point of energy. A violent explosion occurred and the universe began to expand. Let us refer to the time of this singularity as time zero (the beginning of the universe or at least this phase of it). At time zero the universe had an infinite density, an infinite temperature and an infinitesimal extension. The energy of the universe was finite since by conservation of energy, the energy then must be the same as it is now. The present universe is estimated to have a total energy of $10^{69}$ Joules based on an estimate of the total number of stars and galaxies in the observable universe. A description of the universe for times less than $10^{-44}$ seconds when the density of the universe is $10^{94}$ grams per cubic centimeter and the temperature is $10^{33}$ K is impossible because of quantum fluctuations and the uncertainty principle.

## The Hadron Era

It is believed that the universe began as pure light. "And God said: 'Let there be light' And there was light. And God saw that the light was good." (Genesis 1). The light immediately was converted by the mechanism of pair creation into a sea of hadrons, leptons, antihadrons, antileptons and photons. Because of the high energies involved, this fluid of particles was dominated by hadrons. Hadrons continued to be the dominant constituents until the universe expanded to a radius of 30 kilometers after $10^{-4}$ seconds had passed by which time the temperature of the universe had dropped to $10^{12}$ degrees and the density was only $10^{14}$ grams per cubic centimeter. At this point the hadron era came to an end because for temperatures below $10^{12}$ K photons can no longer make hadron-antihadron pairs. The existing hadron-antihadron pairs annihilated each other leaving the $10^{80}$ nucleons, which presently constitute the universe. The reader might wonder how the term era can be applied to the hadron era since it only loses $10^{-4}$ seconds. This instant of time hardly seems to be an era. I remind my readers, however, that a nuclear reaction only takes $10^{-23}$ seconds so there is sufficient time during the hadron era for $10^{19}$ separate reactions to take place for each particle.

There are actually two different descriptions of the hadron era. In the more widely held version it is postulated that the number of baryons (nucleons and other heavy particles) is slightly greater than the number of antibaryons by one part in $10^9$. Therefore, after the annihilation of the hadron and antihadrons pairs takes place only nucleons remained in the universe and no antinucleons. No explanation is given for the initial imbalance of baryons and antibaryons other than to describe the asymmetry as a quantum fluctuation. This version leads to a universe composed totally of matter.

In the second version the number of nucleons and antinucleons is exactly the same. When the universe expands out of the hadron era pockets of matter and antimatter collect due to fluctuations of the density. A pocket of pure matter or pure antimatter is the only possible stable structures that can form since a mixture of matter and antimatter would destroy itself by pair annihilation. This version of the Big Bang theory predicts the existence of two types of clusters, one type composed of matter and a second type composed of antimatter. The number of clusters just equals the number of anticlusters. If a collision were to occur between a cluster and an anticluster, enormous amounts of energy would be released. Nothing of this nature has been observed, however.

On the basis of what we presently know of the universe there is no way of distinguishing between these two versions. Clusters and anticlusters are optically identical. The only way to differentiate them is to observe their interaction with matter. A rocket ship fired from Earth into a cluster would be stable whereas a rocket ship fired into an anticluster would be annihilated. This is an impossible experiment since the nearest clusters are millions of light years away.

## The Lepton Era

As the universe passes out of the hadron era $10^{-4}$ seconds after its inception it passes into the lepton era in which lepton (electrons, muons and neutrinos) and antileptons dominate the landscape. This condition holds until the universe is 10 seconds old, has dropped in temperature to $10^{10}$ K, had achieved a radius of approximately a million kilometers and a density of $10^4$ grams per cubic centimeter. At this point, the photons, which are still in great abundance, can no longer produce electron-antielectron pairs and hence, the lepton-antilepton pairs annihilate each other. According to the first variation of the Big Bang theory, the universe also begins with more electrons than positrons. In fact the electron excess just equals the proton excess so that the number of electrons, which survive the lepton era just, equal the number of protons that survive the hadron era. A very tidy coincidence indeed!

According to the second variation the number of electrons and positrons are the same. The electron and positrons are naturally attracted to blobs of protons and antiprotons during the lepton era because of their mutual electrostatic attraction. The blobs of matter and antimatter are, therefore, electrically neutral like the cluster and anticluster into which they evolve.

## The Radiation Era

The next era the universe passes through is known as the radiation era in which photons and neutrinos illuminate the world completely dominating the protons, neutrons and electrons that have survived the hadron and lepton eras. Matter is ionized into positive and negative charges during this era since the photons have enough energy to ionize any electrically neutral atoms, which might form. While the photons have enough energy to keep electrons separated from nuclei they do not have enough energy to break up deuterium or helium nuclei. The temperatures during the radiation era range from $10^{10}$ K to 3000 K.

The high temperatures that exist during the first half hour of the radiation era bring about thermonuclear reactions, which lead to the formation of helium nuclei principally, as well as the nuclei of deuterium, carbon, oxygen, nitrogen and heavy elements such as iron and titanium. Calculation of the relative abundance of helium and hydrogen due to helium formation during the radiation era agrees exactly with the abundance of helium presently found in the universe both in stars and interstellar space. This is one of the great successes of the Big Bang theory.

The radiation era came to an end after 1 million years when the temperature had dropped to 3000 K. Below this temperature photons no longer have enough energy to ionize atoms. Electrons and protons could finally unite to form neutral hydrogen atoms. During the radiation era neutral hydrogen atoms would form but their lifetime was extremely short because the probability of ionization by a photon was so high. Once neutral matter formed the radiation field became decoupled from matter. It was the beginning of the stellar era.

**The Stellar Era**

The universe at this time was one million light years across and had a density of $10^{-21}$ grams per cubic centimeter, which is approximately 1000 times the present density of our galaxy. At this point, matter began to distribute itself into clusters, galaxies and stars. Today, some 13.7 billion years later, we are still in the stellar era. The universe has expanded to a radius of 13.7 billion light years corresponding to a density of $10^{-30}$ g/cm. The temperature of the photons has dropped from 3000 K to 2.725 K.

This radiation predicted by the Big Bang theory has been observed as an almost isotropic or uniform flux of microwave photons whose average energy is 2.725 K. This result is an important confirmation of the Big Bang theory. In addition to the photons, it is also believed that there exists equal numbers of neutrinos and anti-neutrinos also homogeneously and isotropically distributed. Because of the difficulty of detecting neutrinos this prediction of the Big Bang theory has not yet been verified. The total number of photons and neutrinos is very large, approximately $10^9$ for each nucleon in the universe or $10^{90}$ particles altogether. The energy per particle is quite small but the overall energy content of all the radiation fields is quite large, approximately 1/7 the total energy of the non-dark universe. This radiations field does not interact with matter. It is spread evenly throughout the whole universe, a remnant of the initial explosion, which set the universe on its present course.

Before turning to the nature of clusters, galaxies and star formation in the next chapter we must first examine the role of dark energy in explaining the accelerating expansion of the universe and the role of dark matter in the transition of the universe from a homogeneous fluid into its current state of lumpiness consisting of superclusters, clusters, galaxies, nebulae and stars.

## Dark Energy and the Accelerating Universe

When the Big Bang theory was first proposed it was believed that the universe had been expanding and was expanding at the same constant rate. In 1998 this all changed when the observation of a Type 1a supernovae indicated that the universe was expanding at an accelerating rate. A Type 1a supernova is a white dwarf star made up of carbon that suddenly resumes the fusion process when its temperature increases above the threshold due to the accretion and gravitational condensation of gas from its surroundings. The supernova acts as a standard candle that allows an accurate determination of the distance to the galaxy in which it resides. The conclusion that the expansion of the universe is accelerating has been since corroborated by a number of other observations including cosmic microwave background (CMB) radiation, a more accurate determination of the age of the universe, improved measurements of highly red shifted supernovae and the x-ray properties of galaxy clusters.

An explanation of the acceleration of the universe's expansion requires that much of the energy in the universe consist of a component with large negative pressure, which is identified as "dark energy". Dark energy is also used to explain that the Universe is very nearly spatially flat, and therefore according to General Relativity the Universe must have a critical density of mass/energy. But from the observed gravitational clustering of the observable mass of the universe a great chunk of matter/energy is missing that is needed to explain the nearly spatial flatness of the universe. The missing mass/energy is believed to be made up of dark energy, which is believed to permeate all of the space in the universe. The exact nature of dark energy is not understood. Some claim its is linked to the cosmological constant $\Lambda$, a parameter in Einstein's General Theory of Relativity that we spoke of earlier in this chapter. Others suggest that it is due to a scalar field that cannot be directly observed. Dark energy is believed to make up 72% of the universe with remainder being dark matter which we will describe in the next section making 23% of the universe and the observable matter made up of protons, neutrons, electrons and the other elementary particles that we see in our labs everyday making up a mere 5% of the Universe. This is the challenge facing cosmologists and elementary particle physicists. They only can see 5% of the total universe and from these observations they have to figure out how the other 95% of the universe that they are unable to observe directly behaves and what is the nature of these dark quantities.

## Dark Matter and the Lumpiness of the Universe

The other mysterious substance representing an estimated 23% of the universe is dark matter, which is an electrically neutral form of matter that does, however, exert a gravitational pull or push. Dark matter is a form of matter that is undetectable by electromagnetic radiation but its presence can be inferred from the gravitational effects it has on visible matter. The existence of dark matter was proposed to explain the strength of the gravitational forces within galaxies and between galaxies. The stars in a galaxy rotate about the center of a galaxy where the greatest concentration of stars exists. From the rate of rotation of stars on the periphery of the galaxy and the distance of those stars from the centre of the galaxy one is able to estimate the amount of mass that is required to generate the forces that are observed. In many galaxies there is not enough observable matter in the stars, nebulae and intergalactic gas and dust in the galaxy to account for the strength of the gravitational pull. It is assumed that this deficit is due to dark matter, i.e. matter that cannot be seen because it does not have an electric or nuclear charge.

The existence of dark matter is also required to explain why as a result of the Big Bang the matter we can observe is not uniformly distributed about the universe. Dark matter is responsible for the fact that observable matter gathered together in clumps to form clusters, galaxies and stars. It is suggested that the dark matter, which dominated the early universe amplified tiny inhomogeneities in the distribution of matter causing matter to clump into dense regions and leaving other regions rarefied explaining how the universe went from a homogeneous distribution of energy and matter into one that is lumpy, i.e. one consisting of superclusters, clusters, galaxies, nebulae and stars. The existence of dark matter is required to explain a number of other observations including the lack of uniformity of the cosmic microwave background. Dark matter has to be invoked to explain how clusters of galaxies are able to remain gravitationally bound to each other despite their relatively high velocities. There is not enough observable matter to hold these clusters of galaxies together indicating the existence of dark matter. Finally the phenomenon of gravitational lensing in which the bending of light by a gravitational field has also revealed the existence of dark matter. Although it is quite certain that dark matter exists it is still a mystery as to its actual make up.

According to inflationary theory which explains a number of cosmological features the universe at approximately $10^{-36}$ seconds after the Big Bang underwent an extremely rapid exponential expansion in which its volume increased by a factor of at least $10^{78}$. It is postulated by some cosmologists that the Big Bang that created our observable universe which extends out to 13.7 light years was not the only Big Bang but there have been other and that there are in fact other universes that lie beyond our observable universe. In some versions the laws are the same in these other universes as they are in our universe but the distribution of matter, dark matter and dark energy could be different. And in some versions the laws and the physical constants in the other universe are different from our. The multiverse is the set of all these universes.

# Chapter 26

# Clusters, Galaxies, Black Holes and Stars

## The Super Structure of the Lumpy Universe

Having described dark energy and matter, inflation and the multiverse in the last chapter we can now turn to a description of the visible universe made up of stars, galaxies, clusters and superclusters. There remains for us the task of explaining how the cosmos, at the beginning of the stellar era one million years after the universe itself began, evolved into the presently observed universe of superclusters, clusters, galaxies and stars. The universe in its earliest stages was presumably a uniform sea of matter, which at the beginning of the stellar era had a density of $10^{-21}$ g/cm$^3$. It is known from fluid dynamics that a density fluctuation within a uniform fluid will cause a gravitational condensation of matter. If the density increases in a certain area, then the gravitational forces of the matter within this zone will be stronger than the gravitational pull of matter outside the zone and, hence, the matter within the zone will begin to collapse forming a structure within the uniform fluid. This explains how clusters and galaxies formed from the uniform sea of matter, which composed the universe in the early stages of its existence. The formation of stars within galaxies follows a similar pattern to be described later.

In the last chapter we described the role of dark matter in giving rise to the lumpy structures of our universe today. In this chapter we will describe those structures consisting of stars, galaxies, clusters and superclusters. The stars emerge and live in larger structures consisting of $10^{11}$ stars known as galaxies in which they are held in place by the gravitational pull of the central core of the galaxy. The galaxies themselves are part of a cluster of galaxies that are held in place by their mutual gravitational attraction. Many of these clusters form superclusters where once again gravity binds them together. We begin our story with a description of galaxies beginning with our very own Milky Way galaxy in which our Sun resides after which we will describe a number of different types of galaxies. Once this task is completed we will begin to describe the cluster that contains the Milky Way, the Local Group, and the supercluster, the Virgo Supercluster also known as the Local Supercluster that contains the Local Group and hence the Milky Way. After describing our local cluster and supercluster we will go on to describe the other clusters and superclusters of our universe. We will leave the description of the birth and the evolution of stars to the next chapter.

## The Milky Way Galaxy

The Milky Way Galaxy is a collection of approximately $10^{11}$ stars with a total mass of $10^{41}$ kilograms. The shape of the galaxy is essentially that of a disk with a central spherical bulge. When we view the Milky Way in the night sky we are looking into the plane of the disk. The galaxy appears in the sky as a band of closely clustered clouds, which was given the name the Milky Way and hence became the name of our galaxy. The term galaxy is related to the term Milky Way as the Greek word for milky is γαλαξίας (galaxias).

If viewed from the top of the plane of the disk of stars, one would see that the Milky Way is not a homogeneous distribution of stars, gas and dust, but rather forms a spiral structure and would look very much like the spiral galaxy M81 located in the Ursa Major constellation. The radius of the Milky Way's galactic disk is about 50,000 light years. The Sun is about 26,000 light years from the center located in one of the galaxy's spiral arms. The thickness of the galactic disk varies from 16,000 light years at the center to 6,000 light years at the edge. The very center of the Milky Way galaxy consists of a very large compact object, a supermassive black hole, with a total mass of approximately 4.1 million solar masses. This is typical as most galaxies have a supermassive black hole at their centre.

Radiating out from the centre of the Milky Way galaxy in the plane of the disk are the spiral arms consisting of two major arms and several minor arms. The Sun sits in one of the minor arms. In additions to the spiral arms of the galaxy there are clusters of stars not to be confused with cluster of galaxies that orbit the Milky Way galaxy as satellites. The globular clusters are distributed spherically in the halo of the galaxy beyond the radii of the spiral arms. These clusters are themselves spherical in shape because they are very tightly bound gravitationally due to the high density of stars within them relative to the density of stars in the main disk of the galaxy. Globular clusters contain hundreds of thousand stars, which are much older than the stars in the main disk of the galaxy. To date 158 of these objects have been discovered but it is believed that there might be some others that have not yet been detected. In addition to globular clusters there are open or galactic clusters, which are not tightly bound by gravity and contain only thousands of stars that are relatively new compared to the more numerous stars of the globular clusters. Although open clusters do not contain many stars there are many more of them with well over a thousand having been identified to date but by some estimates could number as many as 10,000. These clusters of stars are formed from clouds of gases and dust and as such are found almost exclusively in spiral and irregular galaxies. One final structure in the Milky Way that needs to be mentioned is that of gaseous nebulae. These vast 'clouds' of gas and dust are considerably denser than the normal interstellar medium of gas and dust and are often regions within the galaxy where new stars are formed. Nebulae appear dark in the sky unless illuminated by a star located either within the cloud or nearby.

The Milky Way Galaxy possesses an enormous amount of angular momentum due to the rotation of the galactic disk about the center of the galaxy. The galaxy does not rotate like a wheel but each star orbits the galactic center in the plane of the disk at its own rate, as is also the case with both globular and open clusters of stars. The period of a star or a cluster depends on its distance from the center of the galaxy. The closer a star is to the center of the galaxy, the shorter is the time for it to complete one revolution just like the planets orbiting the Sun. The reason for this is that the stars are also gravitationally bound to the galactic center and, hence, the radius and the period of their orbit are related by Kepler's laws of planetary motion. The Sun completes a revolution once every 240 million years and, hence, during its 5 billion years existence it has completed a little more than 20 galactic orbits during which time the Earth completed 5 billion solar orbits. Ptolemy and Copernicus were correct: the Earth does move in an epicycle, but one of galactic dimensions.

It is believed that the flattening of the galactic disk occurred as a result of its rotation. The spherical distribution of the globular clusters, which are among the oldest objects in the galaxy, is thought to demarcate the original boundaries of the galaxy.

The age of the Sun is 5 billion years, half the age of the galaxy. The Sun lies in one of the spiral arms where most of the new stars also lie. Fresh material in the form of gas and dust is pouring out of the center of the galaxy traveling along the galaxy's spiral arms with velocities up to 60 kilometers per second. Enough material flows along the arms to produce one star every year. The flow of this material explains the presence of the many young stars observed in the spiral arms.

The spiral arms are also the location of vast magnetic fields. The strength of these fields is only 5 millionths of a gauss which is small compared with the Earth's magnetic field (0.5 gauss in strength). High-energy electrons and protons spiral along the lines of the galactic magnetic field, focusing cosmic rays in our direction and also producing radio waves. These radio signals are called synchrotron radiation because they are produced in almost an identical manner by the man made proton accelerator, the synchrotron.

## The Local Group

The Milky Way Galaxy is surrounded by billions of other galaxies. Those galaxies lying closest to us form a gravitationally bound cluster known as the Local Group, which contains about 20 galaxies altogether. They include our two closest neighbors, the Large and Small Magellanic Clouds, which are two small angular galaxies that are satellites of the Milky Way. The two Magellanic Clouds are 40,000 and 30,000 light years in diameter and have no pronounced structure. Our Local group contains one other large spiral galaxy like the Milky Way. This galaxy has three smaller satellite galaxies bound to it. The remainder of the galaxies of the Local Group are elliptically shaped galaxies of various size, but all are smaller than the two large spiral galaxies. Some of the elliptical galaxies are dwarfs about the same size as a globular cluster.

## Galaxies

We find a remarkable variety in the size and shapes of the galaxies we find within our Local Group. Scanning the remainder of the heavens we discover an even wider range of different types of galaxies. In order to sort out the different types of galaxies, Hubble set up a classification scheme for galaxies based on their shape. He divided the galaxies into four major classes: 1) normal spirals, 2) barred spirals, 3) ellipticals, and 4) irregulars. Normal spirals are galaxies with spherical galactic centers whereas barred spirals have bar-shaped centers. The elliptical galaxies are elliptically-shaped with no spiral arms at all. This class also includes spherical galaxies since the sphere is a degenerate ellipse. The irregular galaxies have amorphous shapes with no distinguishing features. The spiral galaxies are further subdivided according to how tightly coiled the spiral arms are wrapped around the galactic centers. The elliptical galaxies are also further subdivided according to the degree of their ellipticity.

Elliptical galaxies form almost geometrically precise ellipsoids. They rotate in such a manner that their stars orbit the galactic centers in highly eccentric elliptical orbits. The stars of the elliptical galaxies are very old. There is no evidence for new star formation. In the spiral galaxies, on the other hand, new stars are formed in the spiral arms. The irregular galaxies are also rotating systems in which new stars are forming but there is no evidence of spiral arms.

There is a great deal of variation of size within any given class or subclass of galaxies. The largest galaxies belong to the elliptical class of galaxies. Elliptical galaxies can be up to 30 times more massive than the spiral galaxies. They also can be smaller as is the case from some of the elliptical galaxies in our Local Group. The irregular galaxies tend to be smaller than the spirals but here, too, there are exceptions.

Some astronomers believe that the various types of galaxies form an evolutionary sequence so that a single galaxy within its lifetime passes through the various subclasses described above. It has been suggested that irregular galaxies develop spiral arms and evolve into spiral galaxies. The spiral galaxies then evolve through tighter and tighter spiral forms because of their rotation until their arms finally merge into their nuclei. They are now full-fledged elliptical galaxies with a very high degree of eccentricity. At this point the gravitational interaction of the stars in the galaxy produce a more spherical distribution until the galaxy evolves into a perfectly spherical galaxy.

Other astronomers argue that since a large number of elliptical galaxies are 30 times larger than the largest spiral galaxies, it would be impossible for a spiral galaxy to evolve into one of the larger elliptical galaxies. Halton Arp explains the diversity of forms of galaxies in terms of the initial mass and spin of the galaxy. He argues that as the cloud, which formed the proto-galaxy, began to shrink under the force of gravity, the proto-galaxy began to spin more rapidly in order to conserve angular momentum like the figure skater who spins more rapidly as she draws in her arms. If this spin becomes too great, the galaxy will lose mass because of the large centrifugal forces that are generated by the spin. This explains, he claims, why the largest galaxies are elliptical or spherical and not rotating as much as their smaller counter-parts, the spiral and irregular galaxies. The spiral and irregular galaxies were originally high spin objects, which, as a consequence, could not form very massive states because of their spin. Some of the mass they lost might very well have formed the smaller satellite galaxies often associated with these larger high spin galaxies.

**Active Galaxies**

All the objects we have discussed so far are normal galaxies. We shall now draw our attention to a class of objects known as active galaxies. These include radio or exploding galaxies, quasars (quasi-stellar radio sources), and Seyfert galaxies. Active galaxies share a common property, namely, that at their center is an active galactic nucleus, which is a massive black hole with a mass somewhere between $10^6$ and $10^{10}$ Sun masses. This active galactic nucleus radiates much more electromagnetic radiation than any other object in the sky. In fact the massive amount of energy active galaxies radiate cannot be accounted for by the thermonuclear fusion reactions that take place in the Sun and other stars. The energy these objects generate is from the collapse of huge amount of matter from other parts of the active galaxy into the active galactic nucleus, i.e. the massive black hole that sits at the centre of the galaxy. These objects generate some or all of the full range of electromagnetic radiation including radio, infrared, optical, ultraviolet, x-ray and gamma rays. What distinguishes the different members of the class of active galaxies is the mix of radiations they emit. These objects are extremely red shifted indicating that they are very far from us and represent events that took place in the early history of the universe. Radio galaxies consisting of quasars and blazers radiate enormous amounts of radio waves.

The classical mode for the observation of the heavens has always been the detection of visible light, that rather narrow spectrum of electromagnetic radiation that we are able to see with our naked eye. Visible light is not the only type of electromagnetic radiation emitted by the heavens. In fact, the entire range of the spectrum is radiated and we are just beginning to take advantage of this fact. The first type of non-visible sequels to be exploited was radio waves. Work in this area began shortly after World War II, perhaps prodded by the development of radar. A great number of radio sources were discovered. Stars and galaxies that were known from visual sightings were found to be also prodigious emitters of radio waves. In addition to these familiar objects, new objects were discovered that were never observed before. These included quasars and blazers, which form the class of radio galaxies.

## Stars

We live in the stellar era. Aside from dark matter and energy stars are the basic building blocks of our universe. There are approximately $10^{22}$ stars in the universe. They compose 90% of the galactic non-dark material. Stars do not exist in isolation. They are found in galaxies, which provide the concentrations of gas and dust necessary for their formation.

Stars are self-illuminating objects that generate their own energy through thermonuclear fusion. The Sun is a typical star creating its own light and heat. The Sun represents only one of the many types of stars found in the universe. Red giants, white dwarfs, novas, super novae, and pulsars are among the wide variety of stars known to astronomers. The mass of stars range from 0.01 to 50 solar masses where one solar mass is the mass of the Sun or $2 \times 10^{30}$ kilograms (kg). The great variety of stellar types is due partly to stellar evolution and partly to the differences in stellar masses. Stars do not remain in the same state throughout their existence but evolve through a series of stages beginning with their birth as protostars and ending with their death as either black dwarfs, neutron stars or black holes. We shall describe the evolution of stars in this chapter noting the differences due to their masses.

## Star Formation

Star formation takes place in galaxies as a result of the condensation of gas clouds. The presence of dust particles in the gas cloud is essential for this process because the dust serves as a catalyst for the condensation. Once a sufficient amount of condensation has occurred and the density of the cloud is higher than the surrounding interstellar material the cloud will begin to collapse due to the gravitational pull of its own parts. The typical cloud or protostar is a few million light years across. The gravitational collapse due to the gravitational pull of its own parts takes millions of years depending on the mass of the protostar. As the protostar collapses the temperature and pressure within it increases. The collapse stops once the gravitational force is counter balanced by the outward force of the pressure within the cloud. At this point the cloud has achieved stellar dimensions, which are of the order of a million kilometers.

Because of the large extent of the protostar gas cloud and the overall rotation of the galaxy the edge of the cloud away from the galactic center will have a larger velocity than the edge of the cloud closer to the center of the galaxy. As a result the gas cloud has a net rotation or angular momentum. Since the angular momentum or rotational motion of an object is conserved, the rate of rotation of the protostar increases as the cloud shrinks in size. The velocities that develop can become quite large so that if the star retained all its angular momentum it would break apart from the centrifugal forces generated by the spinning motion. Instead there is a gradual loss of material and angular motion as the protostar shrinks in size. Some of the material that is spun off as the protostar shrinks forms the planets and their moons, which also take up a great deal of the angular momentum that is lost.

Stellar material is lost through solar winds and/or the formation of planets. It should therefore not be a surprise to discover that many stars have been observed to have planets given the mechanism for star formation. Stars with masses similar to the Sun are very likely to have planets. The solar wind is a steady stream of gases lost by the Sun and other stars. The Sun has lost almost all of its angular momentum by both planet formation and the solar wind. The planets and their satellites possess 99% of the angular momentum of the solar system. The Sun has the remaining 1%.

If the protostar is much more massive than the Sun then a single star system cannot form and instead a multiple star cluster will form. Stars are never found with masses greater than 50 solar masses. Multiple star systems are as common as single star systems with two star systems or binaries composing almost half the stars in our galaxy. The formation of binaries is still another mechanism for dissipating angular momentum since more angular momentum is contained in the motion of the two stars about each other than in the rotation of the star about their respective internal axes.

## Protostars

The initial stages of the collapse of the protostars from its original dimensions of light years ($10^{13}$ km) to the size of planetary orbits ($10^8$ km) takes place quite rapidly within a matter of a few years. The contraction then slows down. The gravitational potential energy of the protostar is converted into the kinetic energy of the gas molecules as the contraction proceeds. Half of this energy is converted into the internal energy of the gas raising its temperature and the remainder is radiated away as light. The temperature of the center of the protostar when its dimension are that of a planetary orbit are 100,000 K while its surface temperature is about 2500 K. Convection currents of hot gas flow from the center to the edge of the star. As the star contracts more it becomes too dense for the circulation of hot gases. At this point the energy is transferred to the surface by means of radiation rather than convection.

The temperature and density of the star continues to increase until the central region reaches a temperature of 10 million K at which point thermonuclear fusion takes place in which hydrogen is converted into helium. The age of the star when nuclear ignition takes place depends upon its mass. The higher the mass of the protostar the sooner it reaches the temperature and density necessary for nuclear ignition. The Sun, an average size star, spent 27 million years in its pre-nuclear ignition phase, 10 million years in the convection stage and 17 million years in the radiation stage.

## Main Sequence Stars

Once nuclear ignition has taken place the star quickly establishes an equilibrium state in which it remains for the majority of its life. Once the star exhausts its supply of hydrogen gas that has been converted into helium it begins to evolve again. During the long stable period of its existence while it is burning hydrogen, the temperature and luminosity of the star remain fixed. (Note to the reader: I am using terms like ignition and burning to describe thermonuclear fusion but these terms are being used metaphorically because technically the terms ignite and burn are used to refer to oxidation or fire but they are also a handy way to describe fusion.) The force of gravity pulling the star in is just balanced by the outward force of the pressure generated by the dissipation of energy produced by the thermonuclear fusion of hydrogen through the radiation of light.

This balance is extremely stable. If the gravitational force begins to overpower the outward force the star begins to collapse and its temperature and density increase. The rate of nuclear burning increases as a result and more energy is generated and dissipated. The outward force, therefore, increases preventing the star from collapsing any further. If, on the other hand, the inward force begins to overpower the gravitational force the star expands and cools. The rate of nuclear fusion slows down, less energy is dissipated and the outward force decreases by itself. This continues until the outward force just equals the gravitational pull inwards. The star, therefore, remains in stable equilibrium as long as its supply of hydrogen to be converted into helium lasts.

The lifetime of the star in this state depends on its supply of nuclear fuel and the rate at which it uses the fuel. The larger stars, despite their greater fuel supplies, have shorter lifetimes because of the rapid rate at which they burn nuclear fuel. The absolute brightness and the temperature of a star are determined by its mass. The brightness of a star is not determined solely by its temperature but depends on the radius of the star as well. The larger the surface area of a star the more light it will radiate. Temperature and brightness are independent variables. If a two-dimensional plot of the temperature and the brightness of different stars is made it is found that the majority of stars lie along the same line called the main sequence. These stars are in the stable hydrogen-burning phase of their existence. Their position on the main sequence depends solely on their masses. Those stars not on the main sequence are either in the stage of their existence before they undergo nuclear ignition or else they are in their death throes that occur once their supply of hydrogen fuel has been exhausted.

The rate of energy production by a main sequence star due to thermonuclear fusion is quite enormous. The Sun radiates $4 \times 10^{26}$ Joules per second, which requires the destruction of 4 billion kilograms of matter each second. Since 1% of the hydrogen's mass is lost when it is converted into helium the Sun burns $4 \times 10^{11}$ kilograms of hydrogen fuel every second. At this rate the Sun with $2 \times 10^{30}$ kilograms of material would last 150 billion years. Only 10% of the hydrogen of the Sun can be converted into helium, however, since the very high temperatures needed for thermonuclear fusion only occur at the center of a star, a more realistic estimate of the remaining lifetime of the Sun is 15 billion years. The Sun has already existed in its present form 5 billion years. The lifetime of other stars have been estimated to range from half a million years for the brightest and heaviest stars to 200 billion years for the lightest and faintest stars.

## The Sun

Of all the stars in the heavens the Sun is the best known because of its proximity. The Sun is quite a typical star. It shares many of its features with other stars. The current mass of the Sun is $2 \times 10^{30}$ kg, which is in the center of the range of stellar masses. The radius of the Sun, which is spherical in shape, is 700,000 kilometers, approximately 100 times the radius of the Earth. The density of the Sun is only 1.4 grams per $cm^3$ less than the density of the Earth (5.5 grams per $cm^3$) and only slightly greater than the density of water. The temperature of the Sun at its surface is 6000 K and at its center 13 million K.

The surface of the Sun is observed to rotate from west to east. Unlike the surface of the Earth all of which rotates at the same rate the various parts of the Sun rotate at different rates. The equator completes a rotation once every 25 days. The period of the higher latitudes, however, is greater, increasing with latitude and reaching values as high as 33 days. The rotation of the interior of the Sun is not known. Measurements of the oblateness of the Sun, flattening at the Sun's poles, is greater than that one would expect from the centrifugal forces generated by its observed rotation. This might indicate the interior is rotating at a greater rate than the surface.

The structure of the Sun is quite complicated. The interior of the Sun is divided into a number of zones. At the very center energy is being generated by thermonuclear fusion. Above this zone lies the region where energy is transported toward the surface by radiation. Above the radiation zone lies the convection zone where energy is brought to the surface by convection currents. Pockets of hot gas rise to the surface cool off by radiation becoming heavier and then sink below the surface where they are reheated and once again rise. In stars considerably more massive than the Sun the convection and radiation layers are reversed. A photon takes approximately $10^9$ seconds to journey from the center of the Sun to the surface. In the course of its journey it is absorbed and re-emitted several times its wavelength increasing each time.

The surface of the Sun is not well defined like that of the Earth. There is a smooth transition from the interior of the Sun to its atmosphere. The interior of the Sun is opaque and the atmosphere is, therefore, defined to be that point where the gases become transparent. The surface of the Sun or photosphere is by definition, that layer of the Sun we detect visually.

The atmosphere or the chromosphere extends several thousand kilometers above the surface of the Sun. The temperature of the chromosphere falls with altitude. Then suddenly the temperature of the outer atmosphere increases very rapidly reaching temperatures as high as a million degrees. This rise in temperature occurs when the density of the atmosphere becomes very thin. This very thin region is known as the corona. It serves as the launching point for the solar wind, the steady stream of protons, alpha particles and electrons ejected by the Sun.

The surface of the Sun is mottled and granulated. The centers of the granules are hotter than the edges. They form in sizes to about 1000 kilometers in diameter and are due to the hot gases rising to the surface of the Sun. From time to time there appears on the photosphere dark spots known as sunspots. The sunspots are dark because they are cooler than the surrounding surface. Sunspots are caused by disturbances of the Sun's magnetic field and may be regarded as magnetic storms of the Sun. Sunspot activity passes through very active and very tranquil cycles, which reoccur every 11 years. During the periods of intensive sunspot activities solar prominences and solar flares frequently take place. Solar prominences are tremendous jets of solar material which shoot out of the Sun several thousand kilometers above the surface and then fall back again into the Sun. Solar flares are sudden outbursts of radioactive energy from the surface of the Sun, which are several minutes or hours in length. These flares are also associated with an increase in the activity of the solar winds.

The magnetic properties of the Sun are not peculiar to it alone. Other stars are known to have magnetic fields, considerably stronger than that of the Sun, for those stars with smaller radii. The magnetic properties of the Sun and other stars are detected by studying the spectra of their emitted radiation. Magnetic fields split the energy levels of the atoms, which show up in the radiation they emit.

## The Death of Stars

After the hydrogen fuel at the core of a main sequence star has been exhausted the star undergoes dramatic evolutionary changes, passing into a red giant stage. The exact manner in which this change takes place depends on the mass and the chemical composition of the star in question. The general pattern, however, is more or less the same.

## Variable Stars

Some stars in passing from the main sequence stage to the red giant stage pass through a highly unstable phase in which the luminosity and the radius of the star vary in a periodic fashion. During this stage of periodic variation the interior structure of the star is changing. These variable stars known as Cepheid variables were mentioned earlier because of their use in measuring astronomical distances. The period of variation and the absolute magnitude of these stars are related to each other. The period of variation of the Cepheid variables lies somewhere between 1 day and 5 months. Stars do not remain long in the variable phase of their evolution but pass on rapidly to the red giant phase. Other types of variable stars are also observed. Some are known to correspond to later phases of stellar evolution. The exact role others play in stellar evolution is not yet known. Not all stars that leave the stable main sequence to become red giants pass through a variable phase. This depends on their mass.

## Red Giants

Once the hydrogen burning at the core of a star ceases and energy is no longer being generated the helium core that remains begins to collapse. The outer core of the star expands and becomes cooler. The helium core of a star of one solar mass will shrink to about the size of the Earth. This core containing about 1/10 to ¼ the mass of the star will have a density 100,000 times that of the Earth. The temperatures inside the helium core will eventually become so high that a shell of hydrogen surrounding the helium core will undergo nuclear ignition and the star will once again generate new energy. The luminosity of this star will increase to a value of 1000 times its main sequence luminosity. The radius of its outer envelope increases to about 100 times its main sequence radius.

The helium core continues to contract achieving a temperature eventually of 100 billion K at which point helium begins to undergo thermonuclear fusion forming heavier elements, principally carbon. The temperatures and pressure become so high in the core that shortly after the onset of helium burning an explosion takes place in which the helium core increases in size and the hydrogen envelope decreases in size. The temperature of the helium core drops because of its increase in size and, therefore, the rate of helium burning decreases. The hydrogen burning which continues through all these changes also slows down. The star now settles into its second but shorter-lived stable phase with a temperature and luminosity similar to its original main sequence position. It remains in this state until its helium fuel is exhausted. The evolution of stars more massive than the Sun is identical to that described above except the explosion that increases the size of the helium core does not occur because the core of these stars is rather large to begin with.

Once the star exhausts its helium fuel in its second main sequence phase its core, which is composed basically of carbon, now begins to contract. The stars hydrogen envelope expands once again and the star passes into a red giant phase once again. The carbon core continues to shrink as the envelope increases in size. Eventually the envelope becomes so thinned out that neutral atom formation begins to occur because the number of ionizing collisions is so low. As neutral atoms form they limit photons which heat the envelope and accelerate the rate at which it is expanding which produces more neutral atoms which heats the envelope still more and so on and so forth. The expansion proceeds so rapidly that the envelope leaves the star completely forming a ring or planetary nebulae about the star. The planetary nebulae contain perhaps 20% of the stellar materials. It will continue to expand until it dissipates itself in interstellar space after about 50,000 years. The carbon core left behind continues to contract evolving into a white dwarf star.

## Novae

Once or twice a year somewhere in the heavens the sudden brightening of a star occurs. The star will shine with a luminosity 10,000 times its normal output for a few hours. In the days before the telescope this brightening was thought to be due to the birth of a new star, hence the name novae.

It is now believed that this event is associated with planetary nebulae. Novae occur in binary star systems. It is believed that one of the two stars is very hot white dwarf and the other is a red giant with a planetary nebula. It is theorized that gaseous material from the planetary nebula of the red giant falls upon the surface of the white dwarf causing a thermonuclear explosion and a bright flash of light, which we observe as a novae.

## White Dwarfs

The carbon core of a red giant can evolve into a white dwarf only if its mass is less than 1.4 solar masses. Carbon cores with higher mass evolve into other types of stars. The white dwarf state is the final stage of evolution of a star with mass less than 1.4 solar masses. The star can no longer generate new energy. The star contracts into a degenerate gas, which behaves more like a solid than a gas. The white dwarf reaches a minimum size approximately the same size as the Earth, which it maintains even after it, cools. The density of the star is a million times that of the Sun. The white dwarf has an atmosphere only 100 meters thick. It loses its energy through emission of radiation until it becomes a burned-out cinder or a black dwarf hardly radiating at all. Since the universe is only 10 billion years old a white dwarf created at the very beginning of the universe would still only have cooled down to 3000 K which is the lowest temperature ever observed for a white dwarf.

## Supernovae, Neutron Stars And Pulsars

The final days of more massive stars are not as peaceful as those of the white dwarfs. The more massive stars, "rage against the dying of the light." They end their existence not with a whimper but with a bang, a supernovae bang. Supernovae have been observed with the naked eye three times in the past millennium in our galaxy. Supernovae are observed by astronomers with their telescopes on a quite regular basis in distant galaxies. Supernovae in our galaxy occur on average once every 50 years. The first to be recorded in human history took place in the year 1052. This object, which was so bright it could be seen during the daytime, was recorded by people all over the world. The gas clouds from this great explosion are still visible as the Crab Nebula. The second and third supernovae occurred in 1572 and 1604 and were studied by Tycho

Brahe and Kepler, respectively. It was the appearance of the second supernovae, which helped confirm the Copernican hypothesis that the heavens are not immutable but also subject to change.

The exact nature of a supernovae explosion is the subject of some theoretical speculation since the observation of these events is so rare. The following scenario describing a supernova is believed to provide a fairly accurate portrayal of the most powerful stellar events known in today's universe. The energy released by a supernova is $10^{43}$ Joules rivaled only by exploding galaxies and quasars. When a supernova blows it becomes brighter than the entire galaxy that contains it. If the carbon core of a red giant is massive enough the contraction of the core will produce temperatures in excess of 600 million degrees. If the star is not massive enough these temperatures will not be reached and the core evolves into a white dwarf star. When the core temperature of a massive star reaches 600 million degrees, the carbon will begin to fuse to form higher mass elements such as silicon. Once the carbon fuel is exhausted the core will contract once again reaching still higher temperatures at which point the silicon begins to undergo fusion producing still heavier elements, which, in turn, will contract and ignite once the silicon fuel is exhausted. This process of ignition, and fuel exhaustion, and contraction continues rapidly until the core has been completely transmuted to iron and then it stops.

Elements lighter than iron release energy when they undergo thermonuclear reaction. Iron is different. It absorbs energy when it reacts. The production of elements more massive than iron, therefore, ceases. This explains why the abundancy in the universe of elements heavier than iron are so small. With the cessation of nuclear fusion the iron core continues to shrink to the point where there is literally no space between the nuclei. The core has a radius of 10 to 50 kilometers at this point. The density and temperature of the iron core becomes so great that small amounts of heavier elements are produced. The contraction of the core that results is so rapid that a violent explosion ensues in which half the material of the star is ejected violently into the interstellar medium of the galaxy. This material mixes with the gas and dust clouds from which other stars such as our Sun are formed. This explains the presence of the heavier elements found in the Sun and on the Earth. The very material, which you, my reader, and I are composed, was produced in a supernovae explosion by a massive dying star.

The material ejected by the Crab supernovae is seen as the Crab Nebula, which is expanding at the rate of 10,000 kilometers per second or 1/30 the velocity of light. At the center of the Crab Nebula the remains of the exploded core have also been found. This tiny object sends pulses of light and radio waves towards Earth every 0.03 seconds. Other pulsars have been discovered in the heavens with periods ranging from 0.03 to 1.5 seconds.

Pulsars as we will see are neutron stars that are rapidly rotating. According to theoretical consideration stars more massive than the 1.4 solar mass limit for white dwarfs can collapse into neutron stars. The pressures, when stars more massive than white dwarfs collapse, cause the protons and electrons to combine together to form neutrons. The star becomes a degenerate gas of neutrons, which behaves like a solid. The neutron star possesses a very strong magnetic field. The radius of a neutron star is believed to be about 10 to 100 kilometers. This is the same size as a pulsar. The small radius of the neutron star enables it to rotate with extremely high velocities. The neutron star is predicted to emit pulses of light and radio waves with the period of its rotation. The pulses emitted by pulsars have the properties of the pulses a neutron star would produce. Furthermore, calculations show that the remnants of a supernova could easily form a neutron star. It is, therefore, believed that pulsars are nothing more than neutron stars. The reason for the rapid rotation of pulsars and other neutron stars is that as they shrink in size from the stars they descended from their rate of rotation increases to conserve angular momentum, as is the case when ice skaters pull in their arms and rotate at a faster rate.

**Black Holes**

A black hole is an object whose density is so great that space curves back on itself and nothing, not even light, can escape the object. Time comes to a halt within the black hole. There are two basic classes of black holes, namely stellar black holes due to the collapse of a star that no longer has nuclear fuel and supermassive black holes containing many multiples of a solar mass.

The white dwarf and neutron stars are two possible states a star may collapse into at the end of its existence depending on its mass. The collapse of stars to these states occur once the star has exhausted nuclear fuel and can no longer generate the energy necessary to prevent contraction. A third possible form of a collapsed state of a star is a black hole. This state arises only if the star is massive enough, i.e. has a mass greater than 1.4 solar masses. When a black hole forms the gravitational collapse creates a density of matter so great that nothing can prevent the total collapse of the star. The space in which the star is embedded is warped to such a degree that the space closes in upon itself. Once this happens nothing can escape the black hole because the space around it is closed. In other words the gravitational field is so great everything is pulled back to the black hole. This includes light, which is unable to leave the star. This is why a black hole is black. Once an object falls into a black hole it is lost. Nothing can retrieve it. The observation of a black hole presents a problem because it does not send out any information, not even light. The only way of detecting it is to observe the effects of its gravitational field. If it is part of a binary system of stars one can observe matter from the companion star being sucked out of it and falling into the black hole with the consequence of emitting x-rays. One does not observe the black hole directly merely the x-rays from the matter being sucked into the black hole.

Supermassive black holes which contain any where from a thousand to a billion of solar masses form due to the accretion of stellar black holes, normal stars and clouds of gas and dust. They sit at the core of a galaxy, as is the case with our galaxy, the Milky Way. The largest black hole with a mass 18 billion times that of our Sun sits at the core of the active galaxy, OJ 287.

The size of a black hole is determined by its event horizon, which is defined as the radius of the black hole past which any thing that enters never leaves including light. The radius or event horizon of a black hole is determined by its mass. Its radius, known as the Schwarzchild radius, in kilometers, is given approximately by 2.95 times its mass measured in solar masses. The smallest possible black hole would have a mass of 1.4 solar masses and therefore a radius of $2.95 \times 1.4 = 4.1$ km. The largest black hole with a mass of 18 billion solar masses has a radius of $2.95 \times 18 \times 10^9 = 53$ billion kilometers or 350 times the distance between the Earth and the Sun.

Since the radius of a black hole is determined by its mass, it can be described by only three numbers, its mass, its charge and its angular momentum. Any two black holes with the same value of these numbers cannot be distinguished from each other. This is consistent with the notion that that time stops within a black hole since once its mass, charge and angular momentum are fixed nothing can happen.

# Chapter 27

# The Solar System and
# the Planet Earth

The Solar System consists of the Sun and the various bodies, which rotate about it. These include the eight planets (Mercury, Venus, Earth, Mars, Jupiter, Saturn, Uranus and Neptune); the five dwarf planets (Ceres, Pluto, Haumea, Makemake and Eris); the many moons of the planets and dwarf planets; the Asteroid Belt between Mars and Jupiter which also contains Ceres; the Kuiper Belt, which is similar to the Asteroid Belt but lies just outside the orbit of Neptune; comets which can be found in the inner Solar System; the Kuiper Belt; two other trans Neptunian structures, the Scattered Disk and the Oort Cloud to be described later and finally innumerable meteoroids. The distances between these objects are so great that the Solar System consists of a great deal of space, which contains gas, dust and plasma (or ionized gases and electrons). The Earth is 150 million km from the Sun or one astronomical unit (1 AU), which is the unit used to measure distances in the Solar System. The planet farthest from the Sun is Neptune, which is 30.1 AU away. Neptune does not mark the end of the Solar System. The torus-shaped Kuiper Belt extends from 30 AU out to 50 AU and contains Pluto, Haumea and Makemake. The Scattered Disk contains objects including the dwarf planet Eris with highly elliptical orbits that come within 30 AU of the Sun and go as far out as 100 AU. The Oort Cloud extends out to 50,000 AU or one light year and is made up of icy objects and comets.

The planets, dwarf planets and their satellites orbit the Sun in remarkably stable orbits. These nearly circular orbits display a great deal of order and uniformity. The orbits of the planets and their satellites lie in the same plane, which is also the plane in which the Sun rotates about its own axis. The Sun's axis of rotation is perpendicular to the plane of the planetary and lunar orbits. The planets and the moons all orbit in the same direction as the Sun rotates, from west to east with the exception of Venus and Uranus. The planets near the Sun (Mercury, Venus, Earth and Mars) are smaller and denser than the outer planets and are composed of silicates (rocks) and metals. The outer planets of Jupiter, Saturn, Uranus and Neptune are composed primarily of hydrogen and helium. Another distinction between the inner and outer planets is the number of moons. The outer planets have 22 moons altogether while the inner planets have only three moons. The planets and satellites posses 99% of the angular momentum of the Solar System but only 0.1 of 1% of its mass.

Pluto was once considered to be a planet despite its small size when it was first discovered in 1932 but that all changed in 2006 when the term dwarf planet was minted by the International Astronomical Union (IAU) to deal with the controversy of the status of Pluto. As other objects, namely Haumea and Makemake, were discovered orbiting the Sun further out than Pluto but with masses less than that of Pluto the status of Pluto was called into question. With the discovery of Eris, which is larger than Pluto and further out, the controversy of Pluto's status as a planet came to a head. The controversy was resolved by a vote taken at the IAU annual meeting where it was decided that Pluto would be demoted to the status of a dwarf planet and that it would be joined in this category by Haumea, Makemake and Eris. It was also decided that Ceres in the Asteroid Belt would be promoted from an asteroid to a dwarf planet bringing the number of dwarf planets to five for the moment.

Perhaps the most fascinating feature of the Solar System is the spacing of the planetary orbits from the Sun, which follows a pattern known as Bode's law. The radii of the planetary orbits are given by the formula $R = R_0 (0.4 + (0.3) 2^n)$, where n has the values $-\infty$, 1, 2, 3, 4 etc. and $R_0$ is equal to the radius of the Earth's orbit, 150 million km. The values n = $-\infty$, 1, 2, 3, 4, 5, 6, 7 correspond to Mercury, Venus, the Earth, Mars, the Asteroid Belt, Jupiter, Saturn and Uranus respectively. The law does not hold for Neptune, but this planet has a somewhat confused situation because the dwarf planet Pluto is apparently an escaped moon of Neptune. The moons of Jupiter and the moons of Saturn also obey a form of Bode's law.

An explanation of the regularities of the Solar System including Bode's law will require an understanding of how the Solar System formed, which at the moment is not fully understood. It is believed that the planets and their satellites condensed out of the gas cloud that contracted to form the Sun. When discussing star formation we mentioned that as the protostar contracted its rate of rotation increased making it susceptible to losing matter and angular momentum. The matter that the Sun lost during its formation condensed to form the planets and the satellites. A more detailed understanding of the formation of the Solar System unfortunately does not exist. We will, therefore, proceed to describe the various features of the Solar System beginning with the Sun's closest companion the planet Mercury.

## Mercury

Mercury orbits the Sun in a highly elliptical orbit that ranges from 46 to 70 million km (0.31 to 0.39 AU) once every 88 days rotating on its axis once very 59 days. Mercury, the tiniest of the planets, is barely larger than the Moon with a mass only 1/20 that of the Earth. Because of its weak gravitational field the planet has no atmosphere. It resembles the Moon in many ways with its many meteor craters and volcanic mountains. The surface temperature is a scorching 600 K in the day and 200 K in the night. Mercury has no moon.

## Venus

Venus is the brightest object in the sky with the exception of the Sun and the Moon. The planet orbits the Sun every 225 days at a distance of 0.72 AU. Venus rotates on its axis once every 243 days in a retrograde fashion, from east to west instead of the usual west to east manner of all the other rotations of the planets (Uranus excepted) and their moons. The size, mass and density of Venus are very similar to Earth. The planet has no moon, however. Venus has a very thick cloudy atmosphere consisting of carbon dioxide and hydrocarbons, which obscures the features of its surface. The atmosphere is so heavy that the pressure at the surface is 92 atmospheres compared to 1 atmosphere at the surface of the Earth. The surface temperature is surprisingly high, 735 K on the bright side and 550 K on the dark side. These temperatures are attributed to a greenhouse effect. The heavy atmosphere allows solar radiation in but absorbs the radiant energy that is reflected from the surface of the planet.

## The Earth

The planet Earth is a nearly spherical body with a radius of 6,378 km, which is slightly flattened at the poles and bulged at the equator. Its mass is equal to $6 \times 10^{24}$ kg, and its density is 5.4 grams per cubic centimeter or 5.4 times the density of water. The Earth orbits the Sun once every 365 ¼ days or once a year. The extra 1/4 day gives rise to our leap year. The Earth rotates on its axis every 24 hours. The axis of rotation is inclined 23° to the plane of the planetary orbit. The tilting of the axis accounts for the four seasons of the non-equatorial zones of the Earth. When the Earth is tilted such that the northern hemisphere is closer to the Sun it is summer in the North and winter in the South. At the opposite end of the orbit the northern hemisphere is farther from the Sun and the seasons are reversed, winter in the North and summer in the South. In passing between these two extremes the Earth passes through a point where its inclination is perpendicular to the line between the Earth and the Sun. The period corresponds to the transitional seasons, spring and fall.

The Earth possesses an atmosphere consisting primarily of nitrogen and oxygen. Its surface is 71% ocean and 30% land. The surface of the Earth both below the ocean and on the continents displays considerable structure. The interior of the Earth consists of iron-nickel core surrounded by a rocky mantle upon which the crust of the Earth sits. The continental crust is 32 km thick but the crust under the oceans is considerably thinner.

The Earth possesses its own magnetic field, which is related to its rotation as evidenced by the proximity of the geomagnetic North Pole and the geographical North Pole. The Earth's magnetic field has trapped charged particles, which form radiation belts high above the Earth's atmosphere. The Earth has one satellite, the Moon.

## The Moon

The Moon, the second brightest object in the sky, orbits the Earth only 384,000 kilometers away completing a revolution once every 27⅓ days. The Moon always shows the same face to the Earth and, hence, rotates on its axis once every 27.33 days. The Moon is a nearly spherical object with a radius of approximately 1700 kilometers, a little more than 1/4 of the Earth's radius. The Moon's actual shape is egg-like with the longer axis pointing towards the Earth. The mass of the Moon is slightly greater than 1% of the Earth's mass. Despite its small mass and great distance from Earth, its gravitational effect is felt on Earth. The rise and ebb of the tides is due to the gentle pull of the Moon upon the oceans.

The light from the Moon like all the other objects of the Solar System is due to the reflection of the Sun's rays. The waxing and waning of the Moon's phase is due to the fact that different areas are illuminated as the Moon orbits the Earth each month. A new moon occurs when the Moon is between the Earth and the Sun. A full moon occurs when the Earth is between the Sun and the Moon and the Sun fully lights the half of the Moon facing us. From time to time during the time of a new moon the shadow of the Moon is cast upon the Earth causing a solar eclipse. Occasionally at the time of a full moon the Earth casts its shadow on the Moon in which case we have a lunar eclipse.

The Moon has no atmosphere because its gravitational field is too weak to retain any gases. The Moon also has no magnetic field. The Moon appears to be composed primarily of rocks as indicated by its density of 3.3 grams per cubic centimeter. The rocks are basically basalts, which resemble the molten volcanic rock, found on Earth. The Moon apparently formed at the same time as the Earth and passed through a stage in its early history when its surface was covered with molten rock.

The most outstanding feature of the lunar surface is the maria or seas as they were first called because of their resemblance to oceans. These dark patches on the surface of the Moon are responsible for the appearance of the Man in the Moon. In actuality the maria are vast planes strewn with boulders and pock-marked with small meteoric craters. The most dramatic features of the Moon are enormous meteoric craters up to 60 kilometers in diameter. The Moon also has a number of geological features found on Earth such as mountains and volcanic craters.

## Mars

Mars is the best studied of all the planets because of its proximity and transparent atmosphere. It shows a variety of markings, which have fascinated man for centuries. Mars, which appears red to us, is 1.52 AU from the Sun. It orbits the Sun every 687 days and rotates on its axis once every 24 hours and 37 minutes. Its radius is only half that of Earth and its density is less than ours so that its mass is only 11% that of Earth.

The most prominent feature of Mars is its white polar caps, which are due to deposits of frozen carbon dioxide and water. Water has also been found at mid-latitudes in the Martian soil through exploration of its surface. Because of the large quantities of water found on Mars it is possible that it harbours life or it did at one point in its existence. Mars is tilted on its axis and, therefore, undergoes seasonal changes like the Earth. The surface temperature on Mars varies from lows of about $-140°C$ during the polar winters to highs of up to $20°C$ in summers. The polar caps increase during the winter and shrink in the summer. The area between the polar caps is composed of a desert of red dust — the scene of violent dust storms.

The surface of Mars is full of geological features boasting the highest mountains and deepest valleys in all of the Solar System. The most famous surface features of Mars are the Martian canals. The canals, once thought to be the construction of an intelligent being, turns out to be only geological features. The surface of Mars is rich in mountains, canyons and volcanic craters. Martian volcanoes completely dwarf their terrestrial counterparts, rising to heights of 25,000 meters and measuring 500 kilometers across.

Mars' atmosphere is quite thin with an average atmospheric pressure at its surface approximately 1% of that of Earth's. The atmosphere is 95% carbon dioxide, 3% nitrogen, 1.6% argon, with traces of oxygen and water.

Mars has two moons, Phobos and Deimos, which are small and irregularly shaped. Phobos orbits Mars only 9,400 kilometers from the centre of Mars orbiting the planet every 11 hours. Its mass is only $10^{16}$ kilograms or $1.8 \times 10^{-9}$ Earth masses and its radius is only 11.1 km. Deimos has a mass only 1/7th of Phobos but it is a little more than twice the distance to the centre of Mars and orbits Mars every 30 hours.

## Jupiter

Jupiter is the largest planet of the Solar System with a radius of 71,000 kilometers and more mass than all of the other planets taken together. Its mass is 300 times that of Earth's but still only 1/1000 the Sun's mass. The planet orbits the Sun at a distance of 5.2 AU once every 12 years. It rotates upon its axis once every 10 hours giving rise to a centrifugal force resulting in its oblate spheroid shape, i.e. a bulge at its equator and a flattening at its poles. When viewed through a telescope the planet appears to have parallel brown bands on a yellow background. In addition, there is a giant red spot whose dimensions have been as large as 48,000 kilometers by 13,000 kilometers. The giant red spot appears to be floating on the surface of the planet. It is believed to be some kind of storm or disturbance. Its exact nature is a mystery, however.

The planet does not have a clearly defined surface but rather the interior of the planet and the atmosphere flow continuously into each other like the Sun. The planet is composed primarily of hydrogen (71% by mass) and helium (25% by mass) like the Sun with traces of ammonia and methane. It is believed that Jupiter has a rocky core, which is surrounded by the hydrogen and helium gas, which makes up the bulk of the planet. The planet also has an atmosphere made up of hydrogen and helium. The planet resembles the Sun in many ways and is believed to be a star that was not massive enough to ignite itself. The temperature of the surface is approximately 140 K. The planet has a strong magnetic field. It emits synchrotron radiation from its radiation belt, which is considerably more extensive than its terrestrial counterpart.

Jupiter has at least 67 moons. The four innermost satellites of Jupiter are large objects with nearly circular orbits. The other satellites are small and irregular objects and are, very likely, asteroids captured by the giant planet. The planet also has a ring of cosmic dust like Saturn but not as prominent.

## Saturn

Saturn, the second largest planet, orbits the Sun once every 29.5 years at a distance of 9.5 AU. Its mass is 95 times that of Earth. Saturn resembles Jupiter in many ways. It displays the same band structure. It also rotates on its axis once every 10 hours, which accounts for its oblate spheroid shape. It is composed of basically the same material as Jupiter and even has a rocky core, but is less dense and there is a smaller percentage of helium. Saturn emits non-thermal radio waves like Jupiter.

Above the rocky core is thick liquid metallic hydrogen where the distance between the protons is less than the Bohr radius of the hydrogen atom. Above this layer is liquid hydrogen and helium over which a gaseous atmosphere extends for 1000 km. The interior temperature of Saturn is in places as high as 11,400 K like at its core. As a result of this it radiates more than twice as much energy into space as it receives from the Sun.

The most interesting feature of Saturn is its many concentric equatorial rings that stretch from 7,000 km to 80,000 km above Saturn's surface with an estimated local thickness of only 10 to 20 meters. The rings are composed of particles ranging in size from 0.01 to 10 meters and are made primarily of water in the form of ice. In addition to these particles of ice there are tiny moonlets. The mass of all the material in the rings is only about $3 \times 10^{19}$ kg. This is a miniscule fraction of the total mass of Saturn or 0.05 of 1% of the mass of our moon. While only two gaps in the rings can be seen from Earth a fly by of the Voyager spacecraft revealed that the rings are riddled with thousands of gaps some of which are caused by tiny moonlets sweeping out any debris in their orbit. The rings of Saturn have their own atmosphere independent of the planet due to the ultraviolet rays of sunlight interacting with the water to produce molecular oxygen in the form $O_2$.

In addition to the equatorial rings Saturn also has the Phoebe ring, which extends from 128 to 207 times the radius of Saturn and is tilted at 27° to the equatorial plane of Saturn. The rings seems to have been created by tiny meteorites impacting the moon Phoebe which orbits Saturn at a distance of 215 Saturn radii in an orbit tilted at 27° just like the ring.

In addition to the hundreds of moonlets in the rings Saturn has 61 moons. Titan, the largest moon of Saturn, is the second largest moon in the Solar System after Jupiter's moon Ganymede. It exceeds the planet Mercury in size. It is also the only moon in the Solar System that has a significant atmosphere.

## Uranus

Uranus was the first planet discovered with a telescope. It orbits the Sun once every 84 years at a distance of 19.1 AU. It has a radius of 23,800 kilometers. Slightly denser than Jupiter with a mass that is 14.5 times that of Earth. Uranus rotates on its axis once every 10 hours and 50 minutes from east to west like Venus. The equator of most planets is inclined to the plane of their orbits by some 20 or 30 degrees. This accounts for the seasonal change of Earth and Mars. The equator of Uranus, on the other hand, is extremely unusual as it is inclined 97.8 degrees to the plane of its orbit and therefore rotates perpendicular to the line of its orbit about the Sun. The consequence of this is that the poles are in total darkness for 42 Earth years or one half a Uranus year and in light the other half of the planet's orbit.

Uranus differs in composition from Jupiter and Saturn and is more similar to Neptune. Its core is rocky but it also contains ices. Uranus's atmosphere, like that of Jupiter's and Saturn's, is composed primarily of hydrogen and helium but it also contains the ices of water, ammonia and methane along with traces of hydrocarbons. The lower clouds in its atmosphere are made up of water and while those of the upper atmosphere consist of methane. It has the coldest planetary atmosphere in the Solar System, with a temperature as low as 50 K. Winds on Uranus can reach speeds of 900 km per hour.

Uranus has 27 known moons and a complex planetary ring system like that of Saturn. The moons are rather small. The largest Uranian moon has a radius of only 789 km, which is less than half the radius of our Moon. The moons are made up of approximately 50% rock and 50% ice. The 13 known rings are very narrow and are only a few kilometers wide. They are composed of small particles less than a meter in diameter and are thought to be the debris form a moon that disintegrated from a collision of some sort. The rings also contain ice particles.

## Neptune

A search for the planet Neptune was made because of the irregularities of the orbit of Uranus. It was concluded that an eighth planet beyond the orbit of Uranus was causing this perturbation. The position of Neptune was correctly predicted on the basis of the perturbation of the orbit of Uranus. When the telescope was directed at the expected position of Neptune it was discovered in its proper place. Neptune orbits the Sun once every 165 years at a mean distance of 30 AU. It rotates on its axis once every 15 hours and 40 minutes. The planet is very similar to Uranus slightly smaller in size, but a wee bit heavier due to its greater density. Neptune, like Uranus, also appears greenish in colour.

Of Neptune's 13 known moons the largest is Triton. It accounts for more than 99.5 percent of the mass in orbit around Neptune. It is massive enough to be spheroidal unlike the other moons of Neptune. Triton is the only large moon in the Solar System with a retrograde orbit indicating that it was captured rather than formed with the planet. It is likely that it once was a dwarf planet in the Kuiper Belt.

## The Dwarf Planets

There are five known dwarf planets in the Solar Systems, Ceres, in the Asteroid Belt: Pluto, Haumea, and Makemake in the Kuiper Belt and Eris in the Scattered Disk.

**Ceres**, the Solar System's smallest dwarf planet and the only one in the Asteroid Belt, has a radius of 475 km. It has a slightly elliptical orbit with its distance to the Sun varying from 2.58 to 2.99 AU. Its orbital period is 4.6 years. It accounts for one third of the total mass of the Asteroid Belt. It is spherical with a rocky core and an icy mantle of water and hydrated minerals.

**Pluto**, once thought to be a planet was reclassified as a dwarf planet, the second largest in the Solar System with a radius 0.18 that of Earth's and a mass only 0.002 Earth masses. The search for Pluto was motivated by the slight perturbations of Uranus and Neptune. Its orbit is highly inclined and eccentric with its distance to the Sun varying from 30 to 49 AU bringing it within the orbit of Neptune. Its orbital period is 248 years and it apparently rotates on its axis once every 6.39 days. Pluto is composed primarily of rock and ices of water, nitrogen and methane. Pluto has three moons, Charon, the largest and two much smaller moons Nix and Hydra. Charon has a radius one half that of Pluto but only 11.6% of Pluto's mass. Hydra and Nix are tiny objects with diameters of 100 ± 20 km.

**Haumea** has a mass that is one-third the mass of Pluto. It has a rather elliptical orbit with its distance to the Sun varying from 34.7 to 51.5 AU. Its orbital period is 283 years. It has an ellipsoid shape with its long axis twice that of its short axis, which is due to its rapid rotation. Haumea is thought to be composed almost entirely of solid rock and is the remnant of the breakup of a larger object that suffered a catastrophic impact. Haumea has two known moons.

**Makemake**'s orbit is rather elliptical with its distance to the Sun varying from 38.5 to 53 AU. Its orbital period is 310 years. Its diameter is roughly three-quarters that of Pluto and its mass is one-third the mass of Pluto. Makemake has no known satellites.

**Eris** is the largest dwarf planet in the Solar System. Its orbit is very elliptical with its distance to the Sun varying from 37.7 to 97.6 AU. Its orbital period is 557 years. Its diameter is roughly 1.13 times that of Pluto and its mass is 1.27 times the mass of Pluto. Makemake has one known satellite, Dysnomia, whose radius is only 85 ± 35 km.

## Asteroids

Asteroids are defined as small bodies that orbit around the Sun. They are smaller than planets but larger than meteoroids where a meteoroid is an object less than ten meters across making the distinction between an asteroid and a meteoroid quite arbitrary. The largest asteroids like Ceres in the Asteroid Belt and Haumea, Makemake and Eris in the Kuiper Belt have been reclassified as dwarf planets. Again this distinction is again quite arbitrary.

Most of the known rocky asteroids are found in the Asteroid Belt between the orbits of Mars and Jupiter. They are irregularly shaped and made of sulfate rocks. It is estimated that there are between one and two million asteroids larger than 1 km in diameter and millions of smaller ones. The asteroids in the Asteroid Belt rotate with periods ranging from 3 to 20 hours. Over 90% of the asteroids have variable brightness indicating the irregularity of their shape. The irregularity of their shape is likely due to the collision of these objects with each other. A typical example of the irregular shape of the asteroids is provided by Eros, which passes close to the Earth. It is shaped like a brick with a length of 23 kilometers and a thickness and width of 8 kilometers each.

There is a great deal of variation in the orbits of the asteroids. While a majority have nearly circular orbits a large number have highly irregular orbits which brings one asteroid as close to the Sun as Mercury and another as far from the Sun almost as Saturn. At the moment, approximately 1600 asteroids have been discovered and identified. It is estimated that there are as many as 100,000 asteroids altogether. The total mass of all the asteroids in the Asteroid Belt is less than 0.2 of 1% of the Earth's mass, however.

It was originally believed that the asteroids once formed a planet, which broke apart. The total mass of all the asteroids is too low to make this theory credible. Instead it is believed that many asteroids were formed at the same time with dimensions like Ceres and that the fragmentation that has taken place since has been due to frequent collisions.

In addition to the asteroids of the Asteroid Belt there are also objects in the Kuiper Belt that behave very much like asteroids in that they orbit the Sun in nearly circular orbits. They differ from the asteroids in the Asteroid Belt between Mars and Jupiter in that they are made of both rocks and ices of water, methane and ammonia. There is a difference of opinion as to whether these objects are asteroids or not. Some astronomers refer to them as KBOs (Kuiper Belt Objects). Over 1000 have been observed to date but it is estimated that there are many many more. The Kuiper Belt is much larger than the Asteroid Belt with a width 20 times as great and a mass 20 to 200 times greater.

It is believed that the Kuiper Belt formed in the original proto-planetary disc but was too far out from the Sun and therefore there was not enough material to form a planet. Other Kuiper Belt-like structures have been observed around nine other stars.

## Comets

Comets are small bodies usually a kilometer or two in diameter which orbit the Sun in highly elliptical orbits. When they pass close to the Sun they display the tails for which they are famous. More than a thousand comets have been discovered so far, half with the naked eye. Each year 6 to 8 comets appear of which two or three are the return of periodic comets. The majority of comets have orbits, which bring them beyond the orbit of Neptune. Although they most likely have periodic orbits, the length of their periods, which are centuries long, makes it impossible to keep track of them and thus, they appear as one shot objects. A number of comets have shorter periods ranging from five to 100 years. The first periodic comet to be discovered was Halley's comet, which has a period of 76 years. Halley calculated the orbit of a bright comet, which appeared in 1682, noted that its orbit corresponded to the appearance of the bright comets of 1531 and 1607 and predicted the comet would appear again in 1758. It did appear in 1758 and again in 1835, 1910 and 1986 and is due back in 2061. Its appearance was also first reported in 240 BCE.

The orbits of the "periodic" comets differ markedly than the "non-periodic" comets. The periods of the "periodic" comets are naturally much shorter enabling us to observe more than one of their visits. The "periodic" comets orbit in the same direction as the planets and their orbits are more in the plane of the planetary orbits than the non-periodic comets whose orbits are more or less randomly distributed with respect to the planetary plane. The revolutions of the non-periodic comets are evenly divided between direct and retrograde rotation about the Sun.

It is believed that comets form in the debris of matter that lies in either the Scattered Disc or the Oort Cloud. The Scattered Disc contains icy planetoids with highly elliptical orbits that come no closer to the Sun than 30 AU and travel out to distance in excess of 100 AU. The Oort

Cloud is believed to be a spherical region of comets that extends out to 50,000 AU. Although it has never been directly observed it is believed to be the source of the long period comets.

A comet is composed of meteoric material embedded in the ices of methane, ammonia and water. It has been postulated that gravitational perturbation by the planets causes some of these objects to change their stable orbits beyond Neptune into the highly elliptic orbits, which bring the comets close to the Sun.

As a comet passes close to the Sun, the coma and the tail begin to form from the material evaporated by the Sun's radiation. The coma is a spherical shell of gas surrounding the nucleus of the comet. Its size, which is of the order of 100,000 kilometers depends on how close to the Sun the comet approaches — the closer the approach, the larger the coma. The tail, which is frequently curved consists of both gas and dust and grows as the comet approaches the Sun reaching lengths of up to 10 million km. The tail does not follow the comet but rather always faces away from the Sun so that at certain times the comet follows its tail The reason for this is believed to be due to the forces on the gas particles of the tail generated by the solar wind and the radiation pressure of the Sun's light.

During each passage of the comet close to the Sun, it loses approximately 1/2 of 1% of its mass, which means after 200 passages or so the comet disintegrates. The rate of mass loss actually depends on how close the comets pass to the Sun so that it is possible for a comet to break up after only a few passes past the Sun. The distance of the closest approach is usually somewhere between the orbit of Venus and Mars although one comet is known to have come so close to the Sun that it passed through its corona.

## Centaurs

Centaurs are unstable minor planets that lie somewhere between Jupiter and Neptune and have both asteroid and comet like behaviours, hence their name. The first to be discovered was 944 Hidalgo and the largest centaur is 10199 Chariklo with a diameter of 260 km.

## Meteors, Dust and Gas

Meteors are chunks of matter made of stone or metals, which revolve around the Sun. The stony meteors are made of rocks like those found on

Earth. The metallic meteors are 90% iron and 10% nickel with a trace of cobalt. These objects are continuously bombarding the Earth with as many as 90 million arriving each day. As they plunge to Earth they collide with air molecules, which causes them to heat up, and glow. They appear as shooting stars. The heating in the Earth's atmosphere melts the material of the meteor.

Most meteors melt in the atmosphere before they ever reach the Earth. Those that fall to Earth are referred to as meteorites. Hundreds of these objects have been found. The largest stony meteorite found weights 900 kilograms and the largest metallic meteorite weights 31,000 kilograms. Larger meteors than those have fallen to Earth as is evidenced by the many meteor craters, which dot the surface of the Earth. One such crater, the Barrington Crater in Arizona, is 1.28 kilometers across and 174 meters deep. Over 30,000 kilograms of meteoric iron have been gathered at the site. The largest surviving chunk weights 640 kilograms. It is estimated that the meteor that originally caused this crater had a diameter of 60 meters and weighted 900 million kilograms. This meteor is believed to have been a chunk of an asteroid.

Frequently meteor showers occur in which thousands and thousands of meteors descend all at once. This occurs whenever the Earth passes through the debris of a comet or less frequently the tail of a comet. The remains of a comet frequently spread themselves out evenly through the entire orbit of the expired comet. One such swarm of meteors has an orbit, which intersects the Earth's orbit. The Earth passes through this swarm of meteors every year. For two or three weeks around August 12 showers of meteors rain down upon the Earth. Meteors range in size from asteroid chunks down to micro meteors, which are essentially large size dust particles. Dust and meteors probably have a common origin. They are the remains of the original nebular cloud. The interplanetary gas, on the other hand, owes its origins to the solar winds emanating from the Sun.

**Exoplanets or Extrasolar Planets**

We have just reviewed the structure of our own Solar System. We now know that there are many other solar systems out there in space. As of the beginning of 2010 astronomers have observed 405 exoplanets and in a handful of cases two planets have been observed orbiting a single star. It is estimated that there are innumerable numbers of Solar Systems in our galaxy and in the universe as a whole. Most of the exoplanets observed are large gas giants like Jupiter because they are the easiest to observe. Rocky planets like the Earth have been observed and are believed to exist in numbers equal to or greater than that of gas giants. Many astronomers and other scientists believe that many of these planets support life and that there must exist numerous intelligent civilizations out there somewhere in our universe, but because of their vast distances from us we are unaware of them and they of us.

We now return to our Solar System and study in detail the third planet from the Sun, our very own planet Earth.

## The Earth

The third planet from the .Sun the Earth completes a revolution once every 365¼ days and turns upon its axis once every 24 hours. The Earth has a mass of $6 \times 10^{27}$ grams and an average density of 5.4 grams per cubic centimeter. The Earth is almost a perfect sphere with an average diameter of 12735 kilometers. Because the Earth rotates about the axis of its North and South Pole its shape is slightly distorted from that of a perfect sphere. The planet is slightly flat at the poles and bulges at the equator so that the equatorial diameter is 42.8 kilometers greater than the diameter from the North to South pole. This discrepancy of 1/3 of 1% of the Earth's diameter is due to the centrifugal forces generated by the Earth's rotation and indicates that while the Earth is solid it also possesses plastic qualities.

The planet Earth consists of three basic zones corresponding for the most part to the three states of matter, solid, liquid and gas. The principal part of the Earth is the solid globe, which as we will shortly discover consists of the core, the mantle and the crust. There is new evidence that the inner core of the Earth rotates slightly faster than the rocky mantle and crust that cover it by a very small amount which is somewhere between 0.3 to 0.5 degrees per year. The core, which is made of iron, consists of a solid inner core and a liquid outer core. Above the solid Earth lie the oceans, which cover 71% of the Earth's surface. These vast bodies of water have depths up to 10 km. Finally, floating above both the oceans and dry land is the gaseous component of the planet Earth, our atmosphere, which consists basically of nitrogen, oxygen, carbon dioxide, water and traces of other gases.

**The Age of the Earth**

By measuring the age of rocks on the surface of the Earth we are able to determine that the Earth is at least 3.75 billion years old. The oldest rocks found were discovered along the coast of Greenland where they formed some 3.75 billion years ago. Their age was determined using the techniques of radioactive dating. By determining what percent of radioactive krypton has decayed into argon or radioactive uranium into lead the age of the rocks were determined. These results only give a lower limit on the age of the Earth. They essentially tell us how long ago the crust of the Earth formed. It is most certainly likely that the Earth is older than its crust. Measuring the true age of the Earth by looking at features other than the crusts, presents a problem since it is not possible to explore very far below the surface of the Earth's crust.

The way the age of the Earth was determined, oddly enough, was by measuring the lifetimes of meteorites. Making use of the rhubidium-strontium clock one discovers that all meteors were formed at more or less the same time. It was also discovered the ratio of the various isotopes of lead showed a pattern similar to the ratio found in terrestrial rock sample containing different amounts of lead. These patterns indicate that the meteors and the Earth were formed at the same time, approximately 4.6 billion years ago. This age is consistent with the 3.75 billion year age of the crystal rocks.

The Earth formed 4.6 billion years ago but it took almost a billion years for the Earth to cool down to the point where rocks formed on the surface. Other events that took place on Earth can be dated by using radioactive clocks to determine the ages of fossils or the records inscribed in the rocks of mountain formations, or the movement of glaciers during various ice ages. The most notable event of all was the first signs of bacterial life found in rocks over 3 billion years old. The first fossil remains of shell fish and more complicated forms of life are not more than 600 million years old. The first mammals appeared 225 million years ago. The Rocky Mountains formed 65 million years ago, just at the time the dinosaurs were becoming extinct and the first primates were appearing. Two and a half million years ago the latest series of ice ages occurred in which polar and high altitude glaciers advanced and retreated. There have been three other major series of ice ages in the history of the Earth, the first of which dates back to just over 2 billion years ago. We are probably still in the middle of the period since the last ice age terminated only ten thousand years ago. Given the fact of global warming due to the human consumption of fossil fuels it is possible there will not be any more ice ages.

## The Structure of the Earth

The structure of the Earth is not as easy to study as one might imagine. The surface of the Moon, almost half a million km away is easier to study than the interior of the Earth. Man has traveled to the surface of the Moon to collect samples, but has never penetrated more than a few km below the surface of the Earth. It is not an easy matter to dig a very deep hole. Space travel is simpler.

The basic tool for studying the interior structure of the Earth are earthquakes, particularly the propagatum of the shock waves produced by these large scale movements of rocky material within the Earth's crust. Earthquakes release the tremendous amounts of energy generated by the stresses that develop within the crust. Thousands of earthquakes take place every year. Most of these are minor tremors, which release very little energy and cause virtually no damage. Major earthquakes, however, are quite frequent and cause quite a lot of damage. The largest earthquakes release as much energy as a hydrogen bomb and produce in their wake gigantic tidal waves known as tsunamis, which propagate from one side of the Pacific Ocean to the other.

Shock waves from the earthquake propagate through the interior of the Earth. By studying the arrival time of the shock waves at various seismographic stations across the globe one learns of the internal structure of the Earth. It was in this way that geologists discovered that the very center of the Earth is a metallic core surrounded by a rocky mantle upon which the crust is situated. The radius of the Earth is approximately 6400 km. The radius of the iron core on the other hand is only 3500 km, which means the rocky mantle is 2900 km thick. The iron core consists of an inner and outer core. The outer core is molten and the inner core due to the immense pressure is solidified. The material forming the inner and outer cores is probably the same, consisting basically of iron with small amounts of elements such as nickel, cobalt, silicon and sulphur. The density of the core ranges for 9.5 to 13.5 grams per cubic centimeter, the inner core being densest part of the Earth.

The density of the mantle increases with depth from 3 to 6 grams per cubic centimeter. The mantle consists primarily of oxides of silicon, iron and magnesium and contains a number of different layers. The crust whose average thickness is 40 kilometers below the continents and 9 kilometers below the oceans has a density of only 3 grams per cubic centimeter. The crust consists mostly of granite, but layers of sedimentary and volcanic rock are found in most locations as well. It is only on the shields such as the Canadian Shield that the granite base of the crust is exposed.

## The Earth's Magnetic Field

The magnetic field at the surface of the Earth is produced primarily by agents within the interior of the planet. The exact causes are not known although it is popularly believed that the liquid part of the metallic core is responsible for this field. The geomagnetic field is a dipole field identical to the one produced by a bar magnet. The axis of the Earth dipole field makes an angle of 11° with the axis of rotation indicating the magnetic field is associated with the rotation of the Earth.

The Earth's magnetic field at the surface varies in magnitude both with time and with the position on the Earth. The magnetic field strength at the pole is 3 ½ times stronger than at the equator. This is a result of the dipole nature of the internal magnetic field of the Earth.

There are many different types of temporal variations of the geomagnetic field of approximately 1/2 of 1% due to the effects of the Sun, the solar wind, the Van Allen radiation belts of ionized particles and the upper atmosphere. There is also the variation due to the Sun spot activity which also produces a daily variation and which comes in an 11-year cycles. The records of magnetic measurements which have been made at certain locales for the past 350 years indicate that the magnetic field varies by as much as 20% in cycles whose periods are approximately 100 years. Certain anomalous variations of the magnetic field have been observed to drift towards the west at the rate of 1/5 of a degree per year, which indicates a cycle of 2000 years, the length of the time for this anomaly to propagate completely around the Earth. In addition to these rather minor variations on a relatively short term scale the residual magnetism of rocks indicate that the Earth actually reverses the direction of its magnetic field on a time scale of millions of years.

Rocks are composed basically of silicon oxides. They also contain iron oxides, however, which can become magnetized. When a rock is in a molten stage the iron atoms are free to align themselves in the direction of the prevailing magnetic field lines, just as iron filings are aligned by a bar magnet. When the rock congeals the alignment of iron atoms are frozen into the rock. The rock as a consequence, has a residual magnetism indicating the original direction of the magnetic field during its molten state just prior to its solidification. By examining basalts, the rocks formed by volcanoes, or escaping from oceanic ridges one can determine the direction of the Earth's magnetic field at the time of their formation. Sedimentary rocks are also frequently magnetized. Sedimentary rock is formed from deposits of rock grains, which fall to the bottom of the sea. If these rock grains are magnetized then they will align themselves according to the direction of the prevailing magnetic field. When the sedimentary rock forms as a result of the pressure from above these alignments are frozen in and a record of the magnetic field is made.

Evidence for the reversal of the Earth's magnetic field was found in piles of lava flows in which the polarity of the lava changed as one examined deeper and older levels of the lava. These lava piles have been found in different parts of the world. By dating the various levels, using radioactivity techniques, it was found that the polarities of the lava flows from different locales are correlated. The Earth has reversed its magnetic field 25 times in the past 4 million years. The variations have been irregular with millions of years passing between some reversals and only a few thousand years in other cases.

One of the interesting effects of the reversal of the magnetic field is its effect on life. A correlation between the reversal of the magnetic field and the extinction of certain species has been noted. Recent speculation suggests that perhaps the magnetic fields affect the weather and the reversals cause climatic changes, which in turn can cause extinction of certain species. Still another proposal suggests that perhaps the extinct species were directly sensitive to the Earth's magnetic field like certain snails, flatworms and fruit flies living today, and that the reversal of the magnetic field disturbed their life cycle causing extinction.

The mechanism, which causes the reversal of the Earth's magnetic field is not known and is presently a subject of much speculation as is the origin of the Earth's magnetic field. The only theory, which presently can provide a plausible explanation of the Earth's magnetic field and its property of reversing direction, is the dynamo theory. Attributing the Earth's magnetism to the residual magnetism of the rocky mantle could never explain the strength of the Earth's field. Rock lying greater than 30 kilometers below the surface cannot be magnetized any longer because the temperature is too high. It is therefore believed, that the geomagnetic field originates in the core rather than the mantle. It has been suggested that the liquid metallic outer core might be able to produce a field generating electric currents as a result of its fluid motion. The outer liquid metal core is a conductor of electricity. It is known that an electric conductor spinning in an applied magnetic field will begin to conduct a current, which produces its own magnetic field. If energy is provided to the conductor to maintain its spinning motion it will retain its current and associated magnetic field, even if the original applied magnetic field is removed.

This mechanism can explain the geomagnetic field if one can account for the original field to start the process and the present source of energy to maintain the dynamo. The triggering magnetic field could have been produced by electric discharges produced when the Earth first formed. The energy to drive the dynamo has been attributed to radioactivity by some. A more attractive hypothesis is that heat is being generated by the solidification of the inner core, which is slowly increasing its radius. The dynamo theory is very attractive theory for the origin of the Earth's magnetic field because it also provides some hint regarding the origin of polarity reversals as well. Coupled disc dynamos are known to reverse their polarity under certain conditions in a manner not unlike terrestrial polarity flips.

**Origin of the Separation of the Core, Mantle and Crust**

The origins of the basic structure of the Earth, in particular its separation into a metal core, a rocky mantle and a crust, remains as puzzling a mystery as the origin of the geomagnetic field. Two competing theories for the separation of the core from the mantle seem to attract nearly equal support. According to one theory the Earth began its existence as a molten sphere of matter. The heavier material, such as the metal, quickly settled to the center. According to the rival theory the Earth began its existence as a relatively cool planet in a non-molten state. The heat generated by the gravitational accretion of the planet is postulated to have been radiated away initially so that the planet formed in a solid state in which the metals and the rocks are thoroughly mixed. It is alleged that the internal temperature of the Earth has actually increased in certain places since its formation due to the release of energy through radioactivity. This increased temperature has melted the iron but not the rock, which has a higher melting, point. The molten iron has collected in pools and seeped through the rock to the center of the Earth because of its higher density.

The nature and formation of the crust provides less of a mystery than the core mantle separation because of the ease with which the crust may be studied. The Earth's crust floats on the rocky mantle, which is a solid but rather spongy material. The plastic nature of the rocky mantle is due to the heat and pressure to which it is subjected.

The Earth's crust floats on the upper part of the mantle known as the asthenosphere, which lies some 100 to 200 km below the Earth's surface. The crust ranges from 5 to 70 kilometers below the surface with the crust below the oceans extremely thin and composed of basalt or iron magnesium silicate rock. The crust that composes the continents is much thicker, less dense and composed of sodium potassium and aluminium silicate rocks. The continental crust is made up of tectonic plates that move along on top of the asthenosphere. The tectonic plates are less dense and stronger than the asthenosphere upon which they float and move. The movement is driven by the convection of heat and to some degree by gravity and frictional drag. The motion is extremely slow ranging from 10 to 160 millimeters per annum. This motion explains why the continents have separated from the time that all the continents once formed a single land mass. This actually happened twice in the history of the Earth. The first supercontinent, Rodinia, formed one billion years ago. It broken into eight continents some 600 million years and later reassembled to form Pangaea, which with time broke up into eight major plates: 1. The African continental plate, 2. The Antartic continental plate, 3. The Australian continental plate, 4. The Indian subcontinental plate, which includes a part of the Indian Ocean, 5. The Eurasian continental plate, which encompasses both Europe and most of Asia, 6. The North American continental plate, 7. The South American continental plate, and 8. The Pacific oceanic plate.

What drives the separation of the continental tectonic plates is an upwelling of molten rock which spreads out pushing the continents apart. This model has been confirmed by the change of the orientation of the magnetic field of the oceanic crust between the continents. Because the Earth's magnetic filed has flipped over the time period of millions of years the magnetic polarity of the different bands of crust flip back and forth capturing the direction of the Earth's magnetic field over time. This pattern of alternating magnetic polarizations confirms the hypothesis that molten rocks flow out of the Mid-Atlantic Ridge and push the continents apart. It also explains why the east coast of North and South America seem to fit like jig-saw puzzle pieces into the west coast of Europe and Africa. The similarity of rock formations and life forms where these continents would have been in contact provide further evidence supporting this hypothesis. Additional evidence supporting this hypothesis is that the rocks closest to the Mid-Atlantic Ridge are the youngest and those furthest way the oldest.

The places where plates meet are zones of geological activity such as volcanoes, earthquakes and mountain formation. For example the Himalayan mountains are a result of the Indian subcontinental plate colliding with and being thrust under the Eurasian continental plate, which it lifted to create these mountains. Another example is the Pacific Plate's Ring of Fire along the west coast of North and South America, the Pacific east coast of Asia extending down into the south Pacific just to the east of Australia. This zone contains 75% of the active and dormant volcanoes on Earth and is also a very active earthquake zone. The earthquakes occur as a result of the fact that the tectonic plates cannot glide by each other because of friction. The potential energy that builds up as a result of this is released during an earthquake.

## Oceans

The Earth's surface is 71% covered by Oceans or 361 million square kilometers with a volume of 1.3 billion cubic kilometers. The average depth of all the world's ocean is 3,790 meters with a maximum depth is 10,923 meters. The salinity of ocean water is on average about 3.5% salt.

Oceans were the birthplace of life on the Earth. Oceans play an important role in the Earth's climate. They transfer heat through their currents from the tropical zones near the equator to the northern and southern regions of the planet. One example is the Gulf Stream which is like a river running through the Atlantic Ocean from the Gulf of Mexico along the east coast of North America after which is splits in two with the northern stream crossing to Europe moderating the climate there and the southern stream recirculating off West Africa.

Although we have different names for the oceans, Atlantic, Pacific, Indian, Artic and Antarctic there is in fact only one continuous body of saline water covering the Earth's surface encompassing all of the above named oceans and a number of smaller seas such as the Mediterranean, Black, North and Baltic Seas. There are also some saline land-locked-enclosed seas, which do not connect to the world Ocean such as the Caspian Sea, the Dead Sea and the Great Salt Lake.

## Atmosphere

The Earth's atmosphere consists primarily of nitrogen (78%), oxygen (21%), Argon (1%), traces of other gases and up to 1% water depending on the humidity. The atmosphere adheres to Earth because of gravity becoming less and less dense with altitude such that 75% of the atmosphere is within 11 kilometers of the Earth's surface. The boundary between the atmosphere and outer space is not well defined but atmospheric effects on reentering space vehicles begins to be noticeable at about an altitude of 120 km. The total mass of the atmosphere is $5 \times 10^{18}$ kilograms.

One of the components of the atmosphere is the ionosphere, which stretches from 50 to 1,000 km above the Earth and consists of ionized molecules. It has practical importance because it reflects radio waves and permits radio communication across the globe. The ionosphere also gives rise to the northern and southern lights, i.e. the aurora borealis and the aurora australis.

Another component of the atmosphere that is essential for the protection of living organisms from the harmful ultraviolet (UV) rays of the Sun is the ozone layer. Located 10 to 50 kilometers above the Earth this high concentration of ozone ($O_3$) absorbs $95 \pm 2\%$ of the Sun's UV rays. Unfortunately a form of air pollution due to chlorine and bromine fluorocarbons have been breaking down ozone creating holes in the ozone layer over the North and South poles.

Another form of atmospheric pollution is due to the emissions of sulfur dioxide and nitrogen oxides, which give rise to acid rain. Acid rain has a deleterious effect on freshwater and terrestrial ecosystems as well as architectural buildings and monuments such as the Great Sphinx in Egypt.

## Greenhouse Effect

The most serious form of atmospheric pollution, however, is the emission of greenhouse gases, primarily carbon dioxide and to some extent methane. These gases trap sunlight reflected from the Earth's surface and contribute to the warming of the planet. It is agreed upon by the overwhelming majority of the science community that the recent elevation of the level of these gases since the beginning of the Industrial Revolution is due to the activities of humankind. The greenhouse effect is believed to be the cause not only the warming of the planet but also the rapid increase in extreme weather and forest and brush fires.

The scariest possibility is that of a runaway green house effect like the one that occurred on the planet Venus. As our polar caps melt they release trapped carbon dioxide $CO_2$ in the permafrost as well as methane, which is 16 times more effective than $CO_2$ as a green house gas. As the polar caps melt less sunlight is reflected back into space. As the temperature of the Earth and hence the oceans increase the oceans can carry less $CO_2$. Consider how the gas leaves a glass of soda pop as you heat it. These three effects will increase the temperature of the planet melting more of the polar cap and heating the oceans more releasing still more $CO_2$ and methane and one then has the possibility a runaway greenhouse effect that cannot be reversed. Scientists are not sure how much more of a temperature increase would be the tipping point for a runaway greenhouse effect, but for sure we are on the wrong trajectory. We have to take global warming even more seriously to avoid a global catastrophe.

Chapter 28

# Non-Linear Systems, Chaos, Complexity and Emergence

Most systems in nature are inherently nonlinear and can only be described by nonlinear equations, which are difficult to solve in a closed form. Non-linear systems give rise to interesting phenomena such as chaos, complexity, emergence and self-organization. One of the characteristics of non-linear systems is that a small change in the initial conditions can give rise to complex and significant changes throughout the system. This property of a non-linear system such as the weather is known as the butterfly effect where it is purported that a butterfly flapping its wings in Japan can give rise to a tornado in Kansas. This unpredictable behaviour of nonlinear dynamical systems, i.e. its extreme sensitivity to initial conditions, seems to be random and is therefore referred to as chaos. This chaotic and seemingly random behaviour occurs for non-linear deterministic system in which effects can be linked to causes but cannot be predicted ahead of time.

Most of the simple systems that physicists have considered up to the time of the latter half of the twentieth century were simple linear systems giving one the impression that linear systems were the norm and non-linear systems the exception. In fact the opposite is true. Most systems in nature are actually non-linear and chaotic. A system as simple as three bodies problem interacting with each other through gravity is non-linear as was discovered by Poincaré towards the end of the 19th century. He was the first scientist to discover a chaotic deterministic system.

Before the availability of the computing power of the last 50 years the mathematical description of non-linear systems was subject to very limited numerical procedures. With increased computing power, however, scientists have been able to identify new structures and forms of organization within non-linear systems such as fractal structures,

317

emergence and systems of self-organization. They have also been able to better understand the behavior of many non-linear systems ranging from turbulent water flow to volatile stock markets and erratic traffic flows. A self-organizing system is one in which its structure or pattern of behaviour arises entirely from the local interactions of the components of which it is composed. Examples of self-organizing systems in physics ranging from the atomic to the cosmic scale include Bénard cells in fluid dynamics, crystallization, spontaneous magnetization, superconductivity, lasers, star formation and galaxy formation. There are many examples in biology as well including the origin of life itself, the homeostasis of living things and individual cells, bird flocking, fish schooling, the human brain and human culture as in collective intelligence.

Until scientist discovered that complexity and chaos was more the rule than the exception almost all scientists were reductionists in that they believed that complex systems could be understood by reducing them to the simple interactions of their component parts. They believed that the whole is nothing more than the sum of its parts and that eventually one could explain the behaviour of a complex system in terms of the behaviour of its components. Many scientists now acknowledge that one can have systems that are deterministic but whose behaviour cannot be predicted from the behavior of its components. These systems are emergent in the sense that the system as a whole has properties that none of its components possess and it is greater than the sum of its parts. However there is still a large group of scientists who are still reductionists who believe that eventually all phenomena whether biological, psychological or sociological can be explained in terms of basic physics. Proponents of strong artificial intelligence are an example of one such group of reductionists.

Most scientists that accept complexity theory or emergence believe that while one cannot predict the evolution of a complex non-linear system that determinism is still a valid concept. Ilya Prigogine in his book, *The End of Certainty*, takes an even more radical position and claims  not only is it not possible to predict the behaviour of complex system but he claims that the notion of determinism is no longer viable. "The more we know about our universe, the more difficult it becomes to believe in determinism." Prigogine has given up on determinism because of irreversibility of many processes in nature as encompassed in the notion of entropy we studied in Chapter 10. In Newtonian mechanics, which is deterministic the equations describing the motion of bodies can be time reversed and still be valid. In other words a process going forward in time if time reversed would not violate Newton's laws of motion. This means that according to Newton's laws of motion one cannot tell in which direction time is moving. This according to Prigogine represents a denial of the arrow of time and the widely accepted notion that time has only one direction in which it can progress.

Prigogine argues for the irreversibility of time on the basis of many phenomena in the physical world such as radioactivity, weather patterns, the birth and death of living organisms, the origin of life on Earth and its evolution, the evolution of stars, galaxies and the universe itself. We saw that Newtonian physics had to be amended to take into account Special and General Relativity effects through the introduction of 4 dimensional space-time. Classical physics was amended once again to take into account quantum effects and the need for a probabilistic description of individual elementary particles. But although one had to sacrifice the causal description of an individual elementary particle the statistical behaviour of quantum particles can be easily predicted and is in total agreement with empirical observations. Prigogine argues that chaotics and complexity requires a third amendment in that non-linear systems are indeterminate. Those not going quite so far as Prigogine would nevertheless assert that the behaviour of complex chaotics system cannot be predicted to which Prigogine might add, well that makes them indeterminate. Einstein argued that quantum mechanics was incomplete because it had to sacrifice causality saying, "God does not play dice." I wonder if he would argue against Prigogine's position by saying that self-organization is the clever mechanism by which God created this universe of ours.

The focus of this book has been on physics and its description of the physical world. However, embedded within the physical world are living organisms subject to the laws of physics and among those living organisms we find various levels of intelligence including the human brain and/or mind. The brain is also subject to the laws of physics. The descriptions of the world of living things belongs to another science, namely biology and the description of the mental world of human thought belongs to psychology and neuroscience with the former field focusing on behaviour and the latter on the functions of the human brain. But because biological organisms and the brain are subject to the laws of physics and are composed of biomolecules there are some scientists, the reductionists or physicalists, who believe all forms of life and intelligence can be explained in terms of basic physics. When I first began my studies as a physics student I too thought all phenomena could be explained ultimately by physics. Fortunately I grew out of this point of view as I discovered the variety and complexity of the phenomena of my world. I have since adopted a position of strong emergence and believe the evolution of life and intelligence represents emergent phenomena and is indeterminate in the Prigoginian sense described above. Emergence is the phenomena whereby new, unexpected structures arise out of the self-organization of the components of a complex non-linear system. The novel structures or behaviours that arise cannot be predicted from nor derived from the structures or behaviours of the components of the complex system. The key factor in the emergence of emergence (pardon the pun) is the interactions of the components of the system and not just the properties of the individual components.

Philip Clayton (2004) in his book *Mind and Emergence* describes the difference between reductionism and emergence. He identifies three basic schools of thought with respect to the question of the relationship between higher orders of organization such as living organisms and the human brain and the components out of which they are constructed namely biomolecules in the case of living things and neurons in the case of the human brain. The three schools according to Clayton consist of physicalists, dualists and emergentists. The emergentists represent a third option between the physicalists and the dualists according to Clayton. The physicalists believe that all phenomena and all things that exist are basically physical or material and that ultimately everything can be and will be explained in terms of basic physics. The dualists on the other hand believe that in addition to the physical world there is also another element, which is "a soul, self, or spirit that is essentially non-physical (ibid., p. v)." Clayton citing el-Hani and Pereira (2000) describes the emergentist position as consisting of following four elements:

1. All things are made of the basic particles described by physics and their aggregates;
2. As aggregates gain a level of complexity novel properties emerge; These properties cannot be reduced to or predicted from the lower level from which they emerged;
3. Higher-level entities causally affect the lower level entities from which they are composed and from which they emerged in what is called downward causation (ibid., p. 33).

Clayton also identifies two major divisions within the emergence school of thought namely the strong and weak emergentists. Clayton, a strong emergentist himself, as am I, describes strong emergence as the belief that the new higher levels of complexity that emerges are ontologically distinct from the lower levels from which they come and that physics will never be able explain these higher level phenomena. The weak emergence position is that, yes, the levels are distinct but that ultimately they can be reduced to physics once a deeper understanding of the world is achieved. Those that adopt the strong emergence position believe that the properties of the emergent phenomenon cannot be reduced to, derived from or predicted from the components of which they are constructed. The properties of living organisms cannot be reduced to, derived from or predicted from the chemistry of the biomolecules of which they are composed nor can human intelligence be reduced to, derived from or predicted from the biology of the human brain and the nervous system from which it arises. These ideas of strong emergence actually derive from and arise out of the experiences scientists have had with non-linear physics.

## A Comparison of Material and Non-Material Emergence

We now turn to a consideration of the symbolic products of the human mind, namely language and culture. Human symbolic interactions are naturally part of the human biotic system and hence are part of the biosphere. We choose, however, to make a distinction between the purely biological interactions of biosemiosis, on the one hand, and human language and culture, on the other hand. Biosemiosis is the communication of information instantiated in the biomolecules and organs of which living organisms are composed where the information that is communicated is not symbolic, i.e. standing for something else. The sensing of signals from the environment by an organism is another example of biosemiosis. It is therefore the case that the information cannot be separated from those biomolecules or the transmitters or the organs in which they are instantiated. DNA does not symbolize RNA but contributes to its creation chemically through catalysis. The same is true of RNA, it is not a symbol of the proteins it helps to create — it actually catalyzes their chemical composition. The neuronal signals are not symbols of something else but are actual physical signals. The medium and the information content or messages of biosemiosis are the same. Human language and culture, on the other hand, are symbolic in which the information is not instantiated materially but is only physically mediated and as a result are able to move from one medium to another.

We therefore make a distinction between material and non-material emergence. Examples of material emergent phenomena include regular hexagonal convection cells, weather patterns in the abiotic world and living organisms in the biosphere. Non-material emergent phenomena include human language, conceptual thought and culture all of which belong to the symbolosphere. The symbolosphere, originally introduced by Schumann (2003a & b), consists of the human mind and all the products of the mind, namely, its abstract thoughts and symbolic communication processes such as spoken and written language and the other products of the human mind and culture such as music, art, mathematics, science, and technology.

Non-material emergence differs from material emergence in that the first of the four elements el-Hani and Pereira (2000, p. 133) used to describe emergence does not hold, namely that all things are made up of basic particles. Human language, conceptual thought and culture are not made up of basic particles described by physics, they have no extension and they exist in the symbolosphere and not a 6N (where N is the number of particles in the system) dimensional configuration space of physical particles.

As has been argued by Kauffman (2000) and Clayton (2004) that biology cannot be predicted from or reduced to physics. In the same way that biology cannot be reduced to physics it is also the case that the symbolic conceptual non-material aspects of human behavior, namely, language and culture cannot be reduced to, derived from or predicted from the biology of the human brain and the nervous system from which they arise. The symbolic domain of human language and culture are a product of human conceptual thought (Logan 2000, 2006a & 2007) and represent emergent phenomena and propagating organization. They differ from living organisms that populate the biosphere in that they are abstract, conceptual and symbolic and not materially instantiated as such with the exception of technology. In the case of technology it is the concepts and organization that goes into the creation of the physical tools that are emergent and propagate not the actual physical tools.

# Conclusion

Because the living organisms of the biosphere and the conceptual systems of the symbolosphere are emergent phenomena they cannot be reduced to, derived from or predicted from the laws of physics. And so we come to the end of the poetry of physics. Physics cannot help us understand the mysteries of life or intelligence but it can help us understand or at least admire the many fascinating aspects of our universe, a universe that exhibits the fractal structure or self-similarity we talked of above.

## The Fractal Structure of the Universe: An Epilogue

Francesco Sylos Labini of the Enrico Fermi Centre in Rome, Luciano Pietronero of the University of Rome, and Nikolay Vasilyev and Yurij Baryshev of St Petersburg State University argue that cosmological data from the Sloan Digital Sky Survey (SDSS) that encompasses roughly 800,000 galaxies and 100,000 quasars shows that the universe seems to have a fractal structure as far out as our telescopes can see. Their paper published in Nature is disputed by a number of other cosmologists who claim the universe is homogeneous.

But given that fractal structures incorporate self-similarity there many examples at many different scales in which the universe exhibits fractal structures. If we examine living forms on our planet Earth we see many examples of fractal structures of self-similarity. Consider a tree and the way its trunk bifurcates into large branches which then continues to bifurcate into smaller branches, and then into large twig, small twigs and finally leaves. And when we examine the structure of leaves we find another network of bifurcations and the same is true of the root system of the tree. This fractal pattern of self-similarity can be found in the structures of the human body in the circulatory system of the arteries, veins and capillaries and in the fractal branching in the lungs from the trachea, which splits into the two bronchi, which continue to split into smaller and smaller tubes leading to bronchioles, which then lead into the alveoli.

The fractal structure of biological systems has been well established but let us for a moment reflect on the structure of the universe starting with the atom and working our way up to galaxies to see another example of the fractal structures of self-similarity. The basic structure is a tiny component of a system orbiting a nucleus, which has much more mass. We begin with the electron orbiting the nucleus of the atom. But once we jump to large aggregates of matter in space we find a similar structure. Moons orbit planets and planets orbit stars. Stars, which are members of a galaxy orbit the nucleus of the galaxy, which as we have discovered is almost always composed of a supermassive black hole. The sun orbits the center of the Milky Way galaxy every 225 to 250 million years. In addition to the stars orbiting the center of a galaxy there are satellite galaxies that orbit still larger galaxies. The Milky Way galaxy has two smaller satellite galaxies the Large and Small Magellanic Clouds and several dwarf galaxies that orbit it in a structure self-similar to our solar system. Our galaxy belongs to the Cluster of galaxies known as the Local Group with a gravitational center somewhere between the Milky Way and Andromeda galaxies. Finally The Local Group Cluster is part of the Virgo Supercluster, which has two-thirds of its galaxies in an elliptical disk and one-third in a spherical halo. The pattern of smaller objects orbiting a central nuclear mass is repeated over and over again in our restless universe and that, my readers, is still another example of the Poetry of Physics.

Chapter 29

# Classroom Discussions, Activities and Assignments

This book, which was used as a textbook for my Poetry of Physics course, can be read by the individual reader or used as a text for a popular science course. For its use as a textbook I have collected the following suggestions for essay topics or themes for classroom discussions. They can also serve as food for thought for the independent reader.

1. Choose a scientist or a group of scientists from any era discussed in the text and write about how they were a product of their times and/or how they changed or influenced the thinking of their times.

**Examples:** Copernicus, Newton, Einstein, Bohr are examples of scientists who ushered in a new era of science. To what extent was their contribution due to their unique genius and to what extent are they the lucky ones who formulated breakthroughs that eventually would have been discovered? How did their achievements impact on the fine arts, philosophy, religion, or social systems of their times?

2. Analyze a specific scientific or technological development in terms of Thomas Kuhn's (1972) notion of a paradigm shift as articulated in his book *The Structure of Scientific Revolutions*, which was reviewed in Chapter 16.

**Examples:** The heliocentric universe, Galileo's discovery of inertia, Newtonian mechanics, the explanation of heat, the discovery of electric and magnetic forces, Planck's quantization of energy, Einstein's explanation of the photoelectric effect, Einstein's Special and General Theories of Relativity, the Bohr atom, Schrödinger's Wave Mechanics,

Dirac's relativistic quatum mechanics, Feynman diagrams and electrodynamics, Gell Mann's quarks, quantum chromodynamics, the Big Bang theory, dark matter and energy and the idea of the multiverse.

3. Discuss the philosophical implications of some aspect of physics.

**Examples:** Newtonian causality, the breakdown of causality in quantum mechanics, the idea of action at a distance, the concept of entropy, the quantization of energy, the relationship of causality and free will, the expanding universe and the Big Bang, the existence of dark matter and energy. Can all phenomena be reduced to basic physics or is the notion of strong emergence a more appropriate approach to understanding other sciences such as biology?

4. Discuss how science affected the arts, technology, politics, economics or social life or vice-versa how these areas affected science.

**Examples:** The camera obscura and perspective in Renaissance painting; optics, colour theory and Impressionist painting; photography and modern art; Newtonian mechanics and the Industrial Revolution; the printing press and the Science Revolution; the telescope and the Science Revolution. Examine the moral implications of the development of the atomic bomb. Consider this question in light of Michael Frayn's play Copenhagen. Is there a relationship between modern art and relativity and quantum mechanics? Are anarchy and entropy in any way related?

5. Compare a scientific idea or system with the world view or philosophical system of any of the following: a non-scientific culture, a non-Western society, one or several of the major world religions.

**Examples:** Compare Newtonian physics with ancient Chinese, Egyptian, Mesopotamian and Greek physics.

Compare modern cosmology with the Copernican universe or the creation myths of an oral culture.

What are the parallels of modern science and any of the religious texts such as the Bible, the Koran, or the Bhagavad-Gita?

Discuss the implications for religion of the anthropic principle which holds that the constants of nature are not accidental but are fine tuned to make human life possible.

Is there a conflict between religion and science?

Is the Big Bang Theory consistent with one of the opening lines of the Bible: And God said, Let there be light: and there was light (Genesis 1:3) or is this just a coincidence?

If are universe is contained in a multiverse as briefly described at the end of Chapter 25 what are the implications for religion and the notion that God created the universe.

Is scientific truth any more valid than other forms of truth such a mathematical truth?

Does science explain nature or merely describe it? Is there any connection between science and morality?

6. Analyze a piece of literature or a work of art for its scientific content or its use of metaphors from science.

**Examples:** Analyze a poet's use of science metaphor or compare two poets' or a group of poets' use of metaphor such as Donne, Shakespeare, Blake, Goethe, Frost and T.S. Eliot.

Do science fiction writers prepare us for the future or are they writing pure fantasy?

What is the connection between science and the Futurists painters? Discuss my friend the late Leonard Shlain's (1993) book *Art and Physics*: in which he explores the hypothesis that art influenced the development of science.

# Bibliography

Albright, William F. 1957. *From the Stone Age to Christianity, 2nd edn.* Garden City, N. J.: Doubleday Anchor Books

Clayton, Phillip. 2004. *Mind and Emergence: From Quantum to Consciousness.* Oxford: Oxford University Press.

Cottrell, L. 1965. *Quest for Sumer.* New York.

Dryden,, John. 1684. Of Dramatic Poesie.

Eberhard, W. 1957. "The Political Function of Astronomy in Han China." *In Chinese Thought and Institutions.* (ed.) J. Fairbank. Chicago.

El-Hani, Charbel Nino and Antonio Marcos Pereira. 2000. "Higher-level descriptions: why we should preserve them?" in Peter Bogh Andersen, Claus Emmeche, Niels Ole Finnemann and Peder Voetmann Christiansen (eds.) *Downward Causation; Mind, Bodies, and Matter.* Aarhus: Aarhus University Press, pp. 118–42.

Fung, Yu-Ian. 1922. "Why China Has No Science." The International Journal of Ethics, Vol. 32.

Gell Mann, Murray. 1994. *The Quark and the Jaguar.* New York: W. H. Freeman.

Gibb, H. A. R. 1963. *Arabic Literature.* Oxford: Oxford University Press.

Gorini, Rosanna. 2003. "Al-Haytham the Man of Experience. First Steps in the Science of Vision" in International Society for the History of Islamic Medicine. Rome: Institute of Neurosciences, Laboratory of Psychobiology and Psychopharmacology.

Hawkins, Gerald S. 1988. *Stonehenge Decoded.* St. Michael, Barbados: Hippocrene Books.

Hitti, P. K. 1964. *The Arabs.* Chicago: University of Chicago Press.

Innis, Harold. 1972. *Empire and Communication, 2nd edn.* Toronto: Univ. of Toronto Press (1st ed. Oxford, 1950).

Innis, Harold. 1971. *Bias of Communication, 2nd edn.* Toronto: Univ. of Toronto Press (1st edn. Toronto, 1951).

Kauffman, Stuart. 2000. *Investigations.* Oxford: Oxford University Press.

Kramer, S. 1956. *From the Tablets of Sumer.* Indian Hill, Colorado: Falcon Press.

Kramer, S. 1959. *History Begins at Sumer.* Garden City, NY: Doubleday Anchor Books.

Kuhn, T.S. 1972. *The Structure of Scientific Revolutions.* Chicago: Univ. of Chicago Press.

Latourette, K. 1964. *The Chinese: Their History and Culture.* New York.

Levi-Strauss, Claude. 1960. *The Savage Mind.* London.

Lodge, Oliver. 2003. *Pioneers of Science.* Whitefish MT: Kessinger Publishing (1st edn. Oxford, 1893).

Robert K. Logan. 2007. *The Extended Mind: The Emergence of Language, the Human Mind and Culture.* Toronto: Univ. of Toronto Press.

Robert K. Logan. 2004. *The Alphabet Effect.* Cresskill NJ: Hampton Press (1st edn. 1986. New York: Wm. Morrow).

Robert K. Logan. 2003. "Science as a language: The non-probativity theorem and the complementarity of complexity and predictability. "In Daniel McArthur & Cory Mulvihil (eds) *Humanity and the Cosmos*, pp. 63–73.

Long, Charles. 2003. *Alpha: The Myths of Creation.* Oxford: Oxford University Press.

McLuhan, M. 1962. *The Gutenberg Galaxy.* Toronto: Univ. of Toronto Press.

McLuhan, M. 1964. *Understanding Media.* New York: McGraw Hill.

McLuhan, M. and R. K. Logan. 1977. "Alphabet, Mother of Invention." Et Cetera, Vol. 34, pp. 373–383.

Needham, J. 1956. *Science and Civilization.* Cambridge: Cambridge Univ. Press.

Needham, J. 1979. *The Grand Titration.* Toronto: Univ. of Toronto Press.

Neugebauer, O. 1952. *The Exact Sciences in Antiquity.* New York.

Popper, Karl. 1959. *The Logic of Scientific Discovery* (originally written in German as Logik der Forschung). London: Routledge.

Popper, Karl. 1979. *Objective Knowledge: An Evolutionary Approach* (revised edn.). Oxford: Clarendon Press, p. 261.

Schmandt-Besserat, D. 1978. The earliest precursor of writing. Scientific American 238.

Schmandt-Besserat, D. 1980. The envelopes that bear the first writing. Technology and Culture 21, No. 3.

Schmandt-Besserat, D. 1981. Decipherment of the earliest tablets. Science, Vol. 211, No. 3, pp. 283–285.

Schmandt-Besserat, D. 1992. *Before Writing: Vol. 1. From Counting to Cuneiform.* Houston: Univ. of Texas Press.

Schumann, John H. 2003a. The evolution of language: What evolved? Paper presented at the Colloquium on Derek Bickerton's Contributions to Creolistics and Related Fields, The Society for Pidgin and Creole Linguistics Summer Conference, Aug. 14–17, University of Hawaii, Honolulu.

Schumann, John H. 2003b. The evolution of the symbolosphere. Great Ideas in the Social Sciences Lecture, UCLA Center for Governance, Nov. 21.

Shlain, Leonard. 1993. *Art and Physics: Parallel Visions in Space, Time and Light.* New York: Wm. Morrow.

Smolin, Lee. 2006. *The Trouble with Physics: The Rise of String Theory, the Fall of Science and What Comes Next.* New York: Houghton Mifflin Co.